“十四五”高等教育课程思政改革系列教材

制药工程专业英语

主　编　许瑞波　李姣姣

副主编　顾一飞　姚　泳

参　编　袁　君　王洪森

王　燕　黄赛朋

华丽娟　袁　明

南京大学出版社

图书在版编目(CIP)数据

制药工程专业英语/许瑞波,李姣姣主编.—南京:
南京大学出版社, 2024.7
ISBN 978-7-305-27572-2

Ⅰ.①制… Ⅱ.①许… ②李… Ⅲ.①制药工业-化学工程-英语-教材 Ⅳ.①TQ46

中国国家版本馆 CIP 数据核字(2024)第 015956 号

出版发行 南京大学出版社
社　址 南京市汉口路 22 号　　邮　编 210093

书　名 制药工程专业英语
ZHIYAO GONGCHENG ZHUANYE YINGYU
主　编 许瑞波 李姣姣
责任编辑 高司洋

照　排 南京开卷文化传媒有限公司
印　刷 南京鸿图印务有限公司
开　本 787 mm×1092 mm 1/16 印张 18.25 字数 595 千
版　次 2024 年 7 月第 1 版 2024 年 7 月第 1 次印刷
ISBN 978-7-305-27572-2
定　价 58.00 元

网　址:http://www.njupco.com
官方微博:http://weibo.com/njupco
官方微信:njupress
销售咨询热线:025-83594756

前 言

2022 年，中华人民共和国工业和信息化部、国家发展改革委、科技部、商务部、卫生健康委、应急管理部、国家医保局、国家药监局、国家中医药管理局等九部门联合发布《"十四五"医药工业发展规划》，该规划提出了六项具体目标，其一便是国际化发展全面提速，即医药出口额保持增长；中成药"走出去"取得突破；培育一批世界知名品牌；形成一批研发生产全球化布局、国际销售比重高的大型制药企业。围绕该发展目标，提出了"十四五"期间要落实的重点任务，其中最重要的是创造国际竞争新优势——"一是吸引全球医药创新要素向国内集聚，吸引全球创新药品和医疗器械率先在我国注册，提升临床研究国际化水平。二是推动国内医药企业更高水平进入国际市场，支持企业开展创新药国内外同步注册，鼓励疫苗生产企业开展国际认证。三是夯实国际医药合作基础，促进国内外法规接轨、标准互认和质量互信，发挥中药标准全球引领作用，搭建医药国际合作公共服务平台。"完成这些目标和任务的保障之一就是人才建设，不论是新药创新的前沿成果、国内外接轨的行业法规及标准，还是国际合作的同行交流学习，都需要具有制药领域专业英语素养的高素质医药人才。

制药工程是一门由化学、药学、化学工程以及生物工程等交叉结合发展而形成的，研究批量与规模化制造药物以及改变药物的物理力学和生物性质的工程技术学科。制药工程专业英语则是在此基础上针对制药工程专业学生开设的一门课程，在培养学生英语核心素养的同时，提高学生专业相关的能力，使学生能就专业问题在跨文化背景下进行基本沟通、交流及合作，是制药领域人才和企业走向国际化的基本路径。然而，目前市面上制药工程专业英语教材稀少，专业知识涉及面窄，不能做到化学、药学、工程学等多学科专业知识、技能合于一本，且出版年代久远，内容陈旧，脱离现在制药产业的实际情况，不能满足新时代药学人才培养要求。同时，现有教材形态老旧，缺少开放性、弹性和引导性，缺乏思政教育素材；对于专业英语翻译技巧、专业词汇构词法以及撰写专业文献摘要等内容，要么缺乏，要么蜻蜓点水，缺乏系统性及全面性。为了满足社会、行业、企业人才培养需求，满足各高校、研究院对制药工程相关专业英语教学、科研的需求，为制药专业学生及相关领域的从业人员提供一本涉及制药、药学等方面经典的、前沿的专业知识、技能以及专业英语翻译技巧、专业词汇构词法、专业文献摘要撰写方法的参考书，是我们编写本书的目的。

本书的编写遵循教育教学规律，针对学用脱节问题，强化产教融合，邀请国内优秀药企的专家按照"新工科"建设以及当前医药行业对制药专业人才的要求共同编写，增加了教材的实战性、前沿性和可读性。本书共七章，以药物的研发、申报、生产、上市为脉络体系，紧跟药学前沿，内容涉及药物化学、药剂学、药物分析、药理学等多个相关药学学科的最新研究成果，涉及药物研发、申报、生产、上市过程中相关技术、政策、法规等，涉及药品质量控制、临床使用等相关内容，也涉及专业英语翻译技巧、专业词汇构词法以及专业文献摘要撰写方法。满足当前药企、药房、药监部门等各种医药相关行业对制药专业人才的要求。

本书的编写全面落实习近平总书记对教材提出的“五个体现”要求，紧密围绕党和国家事业发展对人才的要求，紧扣加快建设医药强国的战略背景和相关产业发展趋势，以选为主，选编结合，发挥教材绪论、章节内容、拓展材料等各要素的思政内涵承载功能，引导学生在了解当今国内外先进技术及应用现状的基础上，进一步增强服务国家医药事业发展的责任感和使命感。本书在内容和组织结构上进行了精细设计和分析，用爱国、创新、民族复兴、中医药文化、制造强国等思政点贯穿全书，如青蒿素（屠呦呦）、藻酸双酯钠（管华诗）、中药黄芩（《黄帝内经》）等，打造专业知识与思政素材一体化的新颖教材，有利于解决工科学生在综合素质与知识结构方面的缺陷问题，提升人才培养质量。

本书内容注重“以学生为本”，注重教材的学习渐进性，实现从支持“教”到支持“学”的“学材”化转型。不但充分考虑学生的认知特点和学习兴趣，而且考虑到学生以后的发展。内容安排上注重学生的专业素质教育和工程素养培养，广泛结合科技前沿与实践实例，富有特色，激发学生学习兴趣；广泛结合药学专业知识和工程相关知识，将工程理论应用于药品生产，培养学生工程实践能力；广泛结合药学知识和英语知识，在专业英语的教学过程中同时进行药学相关知识的培养。更重要的是把制药专业构词法、翻译技巧、摘要撰写方法编入教材，让学生学会并把这些理论、方法灵活运用到专业英文资料的学习和使用，授之以渔。

本书采用图文并茂的方式、深入浅出，呈现形式多样化的知识传递，非常适合学生课后阅读和学习；以二维码的形式，积极开发补充性、更新性和延伸性教辅资料，激发学生兴趣的同时使学生变被动为主动，将学习延伸至课外，学会应用多种辅助资源，完善学习过程。

参与本书编写的有江苏海洋大学许瑞波、李姣姣、顾一飞；正大天晴药业集团股份有限公司姚泳；淮阴工学院袁君；江苏恒瑞医药股份有限公司王洪森；泰州学院王燕；西北大学黄赛朋；蚌埠医科大学华丽娟、袁明。全书由许瑞波审定、统稿。

特别感谢国家一流专业建设点项目；江苏省“十四五”重点学科建设项目；江苏省重点产业学院建设项目；江苏省产教融合重点基地建设项目；江苏省产教融合型专业建设项目。感谢正大天晴药业集团有限公司、江苏恒瑞医药股份有限公司的大力支持！

在编写本书的过程中，很多新的标准、政策陆续更新发布，且由于编者的经验水平有限，所选内容难免存在纰漏，故书中不足之处还请各位读者谅解并给予批评指正。

编　者

2024 年 6 月

Contents

Chapter 1 Drugs Development

Chapter 2 Drugs Declaration

Chapter 3 Drugs Production

Chapter 4 Drugs Approval Applications and Others

Chapter 5 Translation Techniques for EST

Chapter 6 Specialized English Word-Building Method

Chapter 7 Writing on the Abstract of a Professional Paper

Chapter 1
Drugs Development

Unit 1

Exploration of Artemisinin Derivatives and Synthetic Peroxides in Antimalarial Drug Discovery Research

1. Introduction

Malaria is a devastating infectious disease that is transmitted to humans through the bite of female Anopheles mosquitoes infected with Plasmodium (protozoan parasite). Infrequently, it can also be transmitted by exposure to infected blood products (transfusion malaria) and also through congenital transmission. Five species of Plasmodium (Plasmodium falciparum, P. vivax, P. malariae, P. ovale, and P. knowlesi) are reported to cause malaria in humans. Among these species, P. falciparum and P. vivax are more widespread; with P. falciparum being the most fatal species. P. vivax is mostly found in Latin America and the Indian subcontinent, whereas P. falciparum is the most prevalent malaria parasite especially in the countries of World Health Organization (WHO) Africa region. P. falciparum accounted for 99.7% of estimated malaria cases in 2018. As per the World Malaria Report (2019) released by WHO, malaria caused an estimated 405, 000 deaths and 228 million new cases globally in 2018. The prevention and treatment of malaria mainly rely on chemotherapy and vector control strategies. Vector control is achieved by interrupting the life cycle of malaria parasite, and by creating a barrier between human and mosquitoes. Malaria chemotherapies commonly demonstrate activity against the asexual stages of the parasites and provide supportive therapy to boost the host's immune system. Over the last century, several drugs, including chloroquine (CQ), amodiaquine (AQ), mefloquine (MQ) and artemisinin (ART) have served as mainstay chemotherapeutics against malaria for a long time. Later on different classes of antimalarial drugs including chloroquine, amodiaquine, quinine, pamaquine, primaquine, mepacrine, proguanil, pyrimethamine, sulfadoxine, mefloquine, pyronaridine, halofantrine, hydroxy chloroquine, bulaquine, lumefantrine, ferroquine, isoquine, tefenoquine, piperaquine, tert-butyl isoquine, amopyroquine, BW-A 58C and atovaquone have been developed (Figure 1-1). However, most of the antimalarial drugs as a monotherapy have become useless due to the emergence of drug resistance against P. falciparum. The rising drug-resistance and complicated parasite's life cycle are the major issues which need to clearly understand to completely cure or eradicate malaria the disease.

Currently, there is a pressing need for the development of novel antimalarial leads with outstanding efficacy against drug-resistant and drug-sensitive Plasmodium species. It is pertinent

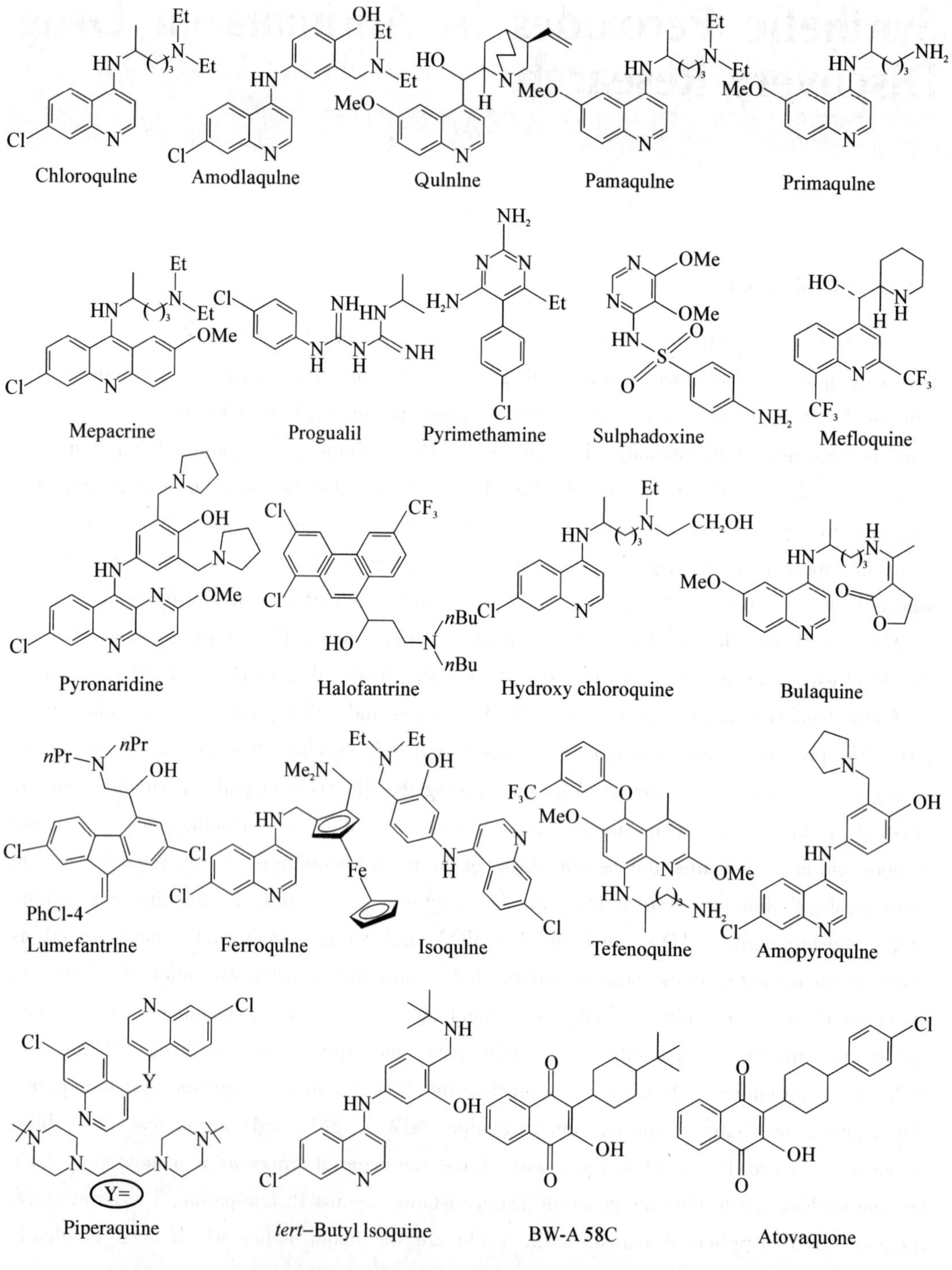

Figure 1-1 Representative structures of different classes of antimalarial drugs

that peroxide based compounds exhibit potential activity against multiple stages of parasite's life cycle with high selectivity and negligible toxicity. In this review, we basically summarize the antimalarial application and successive development of semi-synthetic derivatives (1st and 2nd generation clinical and preclinical candidates) of ART, hybrids of ART with various medicinally valuable structural cores (such as quinoline, ferrocine and quinine etc.), dimers/trimers/tetramers of ART, synthetic peroxides (endoperoxides, ozonides or 1, 2, 4-trioxolane, 1, 2, 4-trioxane, and 1, 2, 4, 5-tetraoxane) based compounds. The compilation of both ART derivatives and synthetic peroxides (non-ART derivatives) in one paper would be very helpful to the medicinal chemists for design and synthesis of novel and simplified structures especially in the field of antimalarial drug discovery.

2. Artemisinin and its derivatives in antimalarial drug discovery

Artemisinin (ART; 1), also referred as Qinghaosu, is a sesquiterpene lactone, first isolated from Artemisia annua L. (family Asteraceae) in 1972. Its molecular structure was assigned in 1979 by a Chinese researcher. ART is a structurally complex natural product which encompasses a tetracyclic core bearing endoperoxide bridge. ART and its derivatives currently form the first line treatment option for malaria. In 2015, the Nobel Prize in Physiology or Medicine was awarded to Chinese pharmaceutical chemist Tu Youyou for her discovery of ART which have saved millions of life worldwide. The ART derived drugs are active against multiple strains of malaria parasite due to their unique molecular structure and also exhibit rapid clearance of malaria parasites from the blood stream as compared to the other available anti-malarial drugs. ART derivatives such as dihydroartemisinin 2 (artenimol), artemether 3, arteether 4, artesunic acid 5 and sodium artesunate 6 have been used for the treatment of malaria. However, these drugs possess low bioavailability and short half-life and are thereby considered as less efficient monotherapy against malaria. Therefore, WHO recommended the use of ART-based combination therapies (ACTs) as a first line treatment for malaria to reduced treatment time, diminished parasite recrudescence and lower probability of resistance development. Inspired by antimalarial potential of ART and its derivatives, several synthetic peroxides based compounds are now in different stages of pre-/clinical development for the treatment of malaria. Apart from the antimalarial activity, ART and its derivatives were also explored for among others, anticancer, antiviral, antiobesity, antidiabetic, antileishmanial, antiinflammatory, anti-bacterial, anti-mycobacterial, and anti-fibrotic activities. These derivatives also demonstrated therapeutic effects against respiratory and kidney diseases. A recent study has shown that ART and its derivatives demonstrate potential anti-severe acute respiratory syndrome coronavirus 2 (anti-SARS-CoV-2) activities *in vitro*. Recently, Pradines and co-workers also reported that a combination of antimalarial mefloquine-artesunate exhibits good inhibitory activity against SARS-CoV-2 replication. These recent findings of ART and its derivatives have been opening a new platform for further research and development to tackle

COVID-19 pandemic. Recently, Babalola and co-workers found that the extracts of Morindamorindiodes root, Morindalucida leaf and Vernonia amygdaline leaf in combination with artesunate reduced parasitaemia by 86.83, 84.20, and 48.10%, respectively against P. berghei NK65 infected mice at doses of 100 mg/kg through oral route. These findings suggests the need for further research on the possible antagonistic, additive, potentiation or synergistic properties of commonly used herbs or natural remedies with antimalarial drugs to combat malaria parasites.

2.1 Brief summary of ART synthesis

The major problem associated with ART and its congener drugs is the poor yield from natural sources (usually >1 wt%) and high cost. Therefore, production of enough amount of ART through isolation from plant remains a challenging task. Several efforts have been made to enhance its production through total synthesis, semisynthetic and flow chemistry pathways in the last 3 to 4 decades. A pioneer report of total synthesis of ART was described by Schmid and Hofheinz in 1983. The authors accomplished the synthesis of ART using (−)-isopulegol in 13 steps with an overall yield of 5%. Later on (1983 – 2004), many other researchers have successfully developed many elegant methods towards the total synthesis of ART. In 1986, Xing-Xiang and co-workers reported the synthesis of ART using (+)-citronellal starting material in 19 steps. In 1987, Avery et al. explored the synthesis of (+)-ART from the 3R-methylcyclohexanone in 12 steps through cyclization of the ozone intermediate. All these synthetic approaches suffered from poor yield and complex procedures. Later on in 1990, a stereo-selective method was described by Ravindranathan et al. for the synthesis of ART through an intramolecular DielseAlder reaction of the (+)-car-3-ene. Another total synthesis method was further reported by Avery and group to synthesize (+)-ART in 10-steps employing (R)-(+)-pulegone starting material. In 1993, Hsing-Jang et al. demonstrated an efficient approach for the synthesis of title compound starting from (−)-β-pinene using an intermolecular Diels-Alder approach. In 1994, Bhonsle et al. explored the total synthesis of ART using (−)-menthol as starting material. Constantino and co-workers described a 10 step synthesis of the abovementioned compound using (−)-isopulegol starting material. Later on, in 2003, a flexible and stereoselective method was developed for the synthesis of ART using an intermolecular radical reaction on iodolactone intermediate and Wittig reaction on the ketone. The use of less number of synthetic steps, high yield and stereo-selectivity as compared to other methods are the key aspects of this approach as explained by Yadav and group. In 2010, the same group reported the synthesis of (+)-ART from R-citronellal as a key substrate. The synthetic protocol involves asymmetric 1,4-addition, Aldol condensation, Ene reaction followed by regio-selective hydroboration as the key steps to synthesize ART. More importantly, this approach is concise protective group free and leads the target compound by shortest route. Another very concise and cost-effective total synthesis of (+)-ART with an overall yield of 9% in 9 steps was reported by Cook and his group in 2012. Despite having valuable development in the field of total synthesis of ART, those approaches are typically too laborious, expensive, and complicated to yield appreciable amount of ART. An effective alternative to reduce the cost is through a semisynthesis from inactive biosynthetic precursors, either artemisinic acid or dihydroartemisinic acid (DHAA), which can be effectively transformed into ART through photochemical processes and thus enhance its production. Artemisinic acid (bicyclic molecule), a quite simple molecular framework, can be easily isolated from the Artemisia annua in higher yields, or produced by fermentation in engineered yeast. Therefore, it could be an ideal substrate to produce ART from synthetic point of view. Besides, DHAA usually derived from artemisinic acid by reduction or isolated by fermentation in engineered yeast. In this context, Seeberger et al. reported two different methods for the semi-synthesis of ART in 39% and 65% yields from DHAA using continuous-flow method. In 2018, Triemer et al. reported a green synthesis of the title compound by reacting crude plant extract (it contains DHAA as the biosynthetic precursor and chlorophyll as a photo-sensitizer), oxygen, acid, and light using continuous-flow process. In 2019, Feng et al. developed heterogeneous dual-catalytic system for the tandem semi-synthesis of ART from DHAA employing Brønsted acid sites into a series of porphyrinic metalorganic frameworks (MOFs). Still, the synthesis of ART from artemisinic acid has proven a challenging task for chemist, since a high-yielding, concise, scalable, and low-cost approach for synthesizing a highly complex molecule is still needed and is a hot-topic of research to continue.

2.2 ART: mode of action

ART, its first-generation derivatives and related endoperoxides exhibit potential antimalarial activity against the blood stages of Plasmodium, which includes the ring stages and mature trophozoites (Figure 1-2). ART derivatives also show some degree of activity against various stages of gametocytes. The endoperoxide linkage of ART derivatives has proven an essential part to display antimalarial efficacy, for instance, deoxoartemisinin (lack endoperoxide bridge) is devoid of efficacy. From the mechanistic point of view, endoperoxides usually undergo reductive activation in presence of heme [Fe(Ⅱ)] (released during parasite Hb digestion) which initiates homolytic fission of peroxide linkage to produce oxygen-centered radical and then carbon centered radical (stable cytotoxic species). These radicals cause cell membrane damage, oxidation of fats and proteins, inhibit the synthesis of nucleic acid and proteins and interaction with cytochrome oxidase and the glutamine transport system in parasites. The mechanism described in Figure 1-3 depicts a typical oxidative process that causes parasite cell death after many cascade events occur in parasite cell. Posner and co-workers demonstrated that endoperoxide linkage of ART is a trigger, activated by iron-induced reduction inside the malaria parasite. This process leads the production of cytotoxic free radicals, one or more high-valent iron-oxo intermediates, and electrophilic alkylating agents that ultimately cause fatal damage to the parasites. Interaction of lipid-solubilized heme with ART, followed by Fe(Ⅱ)-mediated generation of radical species (oxygen and carbon centered radicals) in the vicinity of lipid bilayer, hydrogen radical abstraction, allylic radical generation and subsequent peroxidation by molecular oxygen lead to the formation of lipid hydroperoxides. This process generates reactive oxygen species (ROS) which leads to oxidative stress and subsequent damage of various cellular processes, inhibition of protein and enzyme synthesis and finally parasites' cell death. It has also been demonstrated that the antimalarial efficacy not solely depends on the cleavage of peroxide bond but also on the ability of reactive intermediates to alkylate heme or other proximal targets. It has been suggested that the damage caused to the parasite's vacuolar membrane is due to Fe(Ⅱ)-mediated bioactivation of endoperoxide ring and resulting oxidative stress. It is believed that alkylation (radical species directly target heme and alkylate heme molecule) is another key mechanism of endoperoxide-based antimalarial drugs.

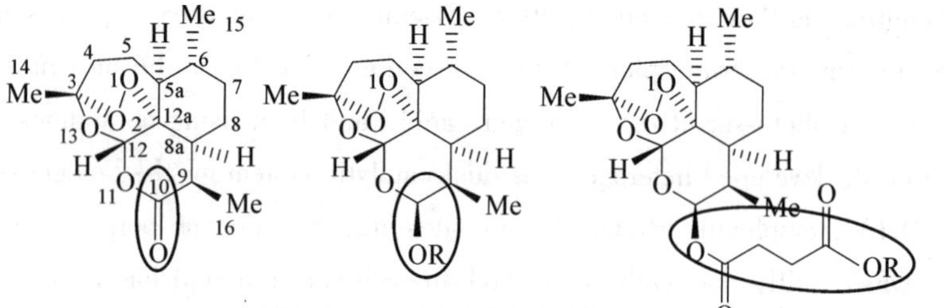

Artemisinin(1)

Dihydroartemisinin(2); R=H
Artemether(3); R=Me
Arteether(4); R=Et

Artesunic acid(5); R=H
Sodium artesunate(6); R=Na

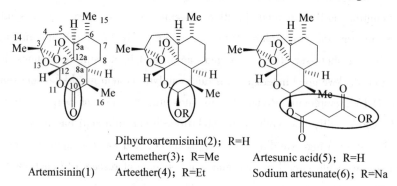

Figure 1-2 Artemisinin and its first-generation derivatives

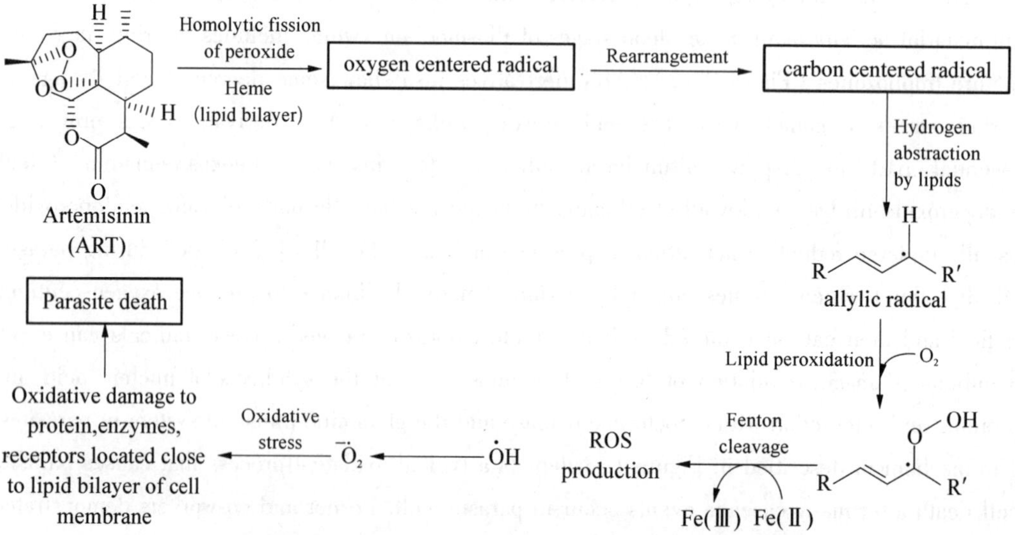

Figure 1-3 Mechanism of action of ART and its derivatives

引自：PATEL O P S, BETECK R M, LEGOABE L J. Exploration of artemisinin derivatives and synthetic peroxides in antimalarial drug discovery research [J]. European Journal of Medicinal Chemistry, 2021, 213, 113193.

Words and Expressions

Malaria	*n.* 疟疾	Plasmodium	*n.* 疟原虫
falciparum	*n.* 恶性疟原虫	chemotherapy	*n.* 化疗
vector	*n.* 载体	immune	*adj.* 免疫的
chloroquine	*n.* 氯喹	amodiaquine	*n.* 阿莫地喹
artemisinin	*n.* 青蒿素	antimalarial	*adj.* 抗疟疾的
parasite	*n.* 寄生虫	drug-sensitive	*adj.* 药物敏感
toxicity	*n.* 毒性	semi-synthetic	*adj.* 半合成
preclinical	*adj.* 临床前的	peroxide	*n.* 过氧化物
lactone	*n.* 内酯	endoperoxide	*adj.* 内过氧化物
anticancer	*adj.* 抗癌的	antiobesity	*adj.* 抗肥胖性的
inflammatory	*adj.* 抗炎的	fibrotic	*adj.* 纤维化的
bacterial	*adj.* 细菌性的	antiviral	*adj.* 抗病毒的
Nobel Prize	诺贝尔奖	natural product	天然产物

Notes

① The rising drug-resistance and complicated parasite's life cycle are the major issues which need to clearly understand to completely cure or eradicate malaria the disease.
耐药性的增加和复杂的寄生虫生命周期是彻底治愈或根除疟疾需要清楚认识的重大问题。

② These drugs possess low bioavailability and short half-life and are thereby considered as less efficient monotherapy against malaria.
这些药物具有较低的生物利用度和较短的半衰期,因此被认为是治疗疟疾的低效单一疗法。

③ This process generates reactive oxygen species (ROS) which leads to oxidative stress and subsequent damage of various cellular processes, inhibition of protein and enzyme synthesis and finally parasites' cell death.
这一过程产生的活性氧,能够导致氧化应激和随后的各种细胞过程的损伤,从而抑制蛋白质和酶的合成,最终导致寄生虫的细胞死亡。

④ SARS-CoV-2
英文全称:Severe Acute Respiratory Syndrome Coronavirus 2。中文全称:新型冠状病毒。

Exercises

1. Put the following into English:

蛋白　　烷基化　　二氢青蒿酸　　产率　　毒性

2. Put the following into Chinese:

a. From the mechanistic point of view, endoperoxides usually undergo reductive activation in presence of heme [Fe(Ⅱ)] (released during parasite Hb digestion) which initiates homolytic fission of peroxide linkage to produce oxygen-centered radical and then carbon centered radical (stable cytotoxic species).

b. This process leads the production of cytotoxic free radicals, one or more high-valent iron-oxo intermediates, and electrophilic alkylating agents that ultimately cause fatal damage to the parasites.

Reading material A:

Tu Youyou: the First Chinese Nobel Laureate

Tu Youyou, a researcher at the China Academy of Chinese Medical Sciences, Chinese pharmacologist, was born in 1930 in Ningbo, eastern China's Zhejiang province. She became the first Chinese person to receive the US-based Lasker Award for clinical medicine in 2011. And she is the first Chinese Nobel laureate in physiology or medicine for the discovery of artemisinin in 2015. She received half of the year's medicine prize of 4 million Swedish krona ($473, 000; 429, 000 euros). The other half was shared equally by Irish-born US scientist William Campbell and Satoshi Omura of Japan, who jointly discovered a novel therapy against infections caused by roundworm parasites. During Tu's stay in Sweden, she donated her personal items, a plate marking her achievements and a book of her study on artemisinin, to the Nobel Museum.

Together with her team, the pharmacologist Prof. Tu derived artemisinin from sweet wormwood, which she found cited in a fourth-century TCM text as an ingredient to cure fever, developing a crucial drug that has significantly reduced mortality rates among malaria patients in recent decades. She has urged more research into the benefits of traditional Chinese medicine.

"The discovery of artemisinin has led to development of a new drug that has saved the lives of millions of people, halving the mortality rate of malaria during the past 15 years," said Professor Hans Forssberg, a member of the Nobel committee for physiology or medicine, when presenting Tu's scientific contributions.

Dressed in purple for the Stockholm ceremony, Tu delivered a speech titled Artemisinin is a Gift from TCM to the World. Several days before, at a news conference at the Karolinska

Institute in Stockholm, Tu called for joint efforts worldwide to combat malaria and develop more potential uses for TCM, which she called a "great treasure" with thousands of years' history and empirical knowledge. She said that by combining TCM with modern scientific technologies, "more potential can be discovered in searching for new drugs".

"Malaria is a pandemic that can go easily out of control, especially in low-income regions such as Africa. So all parties should, under the framework of the (World Health Organization), try their best to delay the process of artemisinin resistance," Tu said.

According to the WHO, more than 240 million people in sub-Saharan Africa have benefited from artemisinin, and more than 1.5 million lives are estimated to have been saved since 2000 thanks to the drug. Apart from its contribution to the global fight against malaria, TCM played a vital role in the deadly outbreak of severe acute respiratory syndrome across China in 2003.

Besides treating viruses, TCM has been most effective in diagnosing diseases; cultivating fitness; treating chronic, difficult, multisource illnesses; and using nonmedicinal methods such as acupuncture and breathing exercises. However, TCM, which is based on a set of beliefs about human biology, including the existence of a life force, qi, is seldom understood or embraced by the West. Some have even labeled it a "pseudoscience", even though it is based on more than 2,000 years of practice in China. Tu's acclaim will bring more recognition and respect for TCM, experts say. The Western world should learn to appreciate the value of the treasures of TCM, which would lead to more basic scientific research into ancient TCM texts and ways to explore research findings worldwide.

Although the traditional way of using TCM treatments has benefited Chinese for millennia, the process highlighted by the Nobel Prize carries its own advantages in the prescription and administration of the medicine and related logistics.

The success has led to a promising approach to the modernization and development of TCM through science and technology. Tu was inspired by many ancient classical works on TCM pointing to sweet wormwood as a candidate for the development of an antimalarial drug. Her success demonstrates that TCM has to embrace modern technologies and laboratory tools and, more important, stick to the essence of time-honored medical science.

As we known, international standards for acupuncture needles and the use of ginseng have been set by China. A simple application of Western medical standards to TCM won't work, so more research and standards for traditional medicine is needed urgently.

The discovery of artemisinin is an excellent example of the development of TCM. With the gradually deepening knowledge of the efficacy of TCM, formulations are constantly changing. Artemisinin is the product of Chinese medicine entering the modern cellular and molecular era. Tu's success may encourage Westerners to learn more about Chinese medicine.

Artemisinin will continue to be the priority drug for the treatment of malaria, despite signs of drug resistance to artemisinin-based combination therapies (ACT) in some countries, Prof.

Tu said, her team has proposed solutions to the problem of artemisinin resistance, adding it was still "the best weapon" against malaria.

"Artemisinin continues to work very well in most of the world, and it will remain highly efficacious," said Pedro Alonso, director of the Global Malaria Programme of the WHO.

Reading material B:

Tu Youyou: China's First Nobel Prize Laureate in Medicine

Unit 2
β-Lactams Past and Present

A cavity in the foundation stone laid for the North West Extension to St Mary's Hospital Medical School in Paddington, London, contains some memorabilia connected with famous employees of the Hospital. One item is a stopwatch recording the time of the first sub-four-minute mile run by Dr Roger Bannister; also contained in the stone is a replica of the culture plate that led Sir Alexander Fleming to the discovery of penicillin. Bannister's achievement was a tribute to his dedication, training and supreme physical fitness, while Fleming's discovery of the first antibiotic is a classic example of a scientist finding an unusual, at the time possibly annoying, outcome in a standard experiment and carefully following up the observation to reveal a major scientific breakthrough.

Isolation from natural sources

It was in September 1928 that fungal spores gained access to Fleming's laboratories (either from the street below or from an adjacent mycobiological laboratory) and settled on an agar plate containing *Staphylococcus* bacteria. In Fleming's own words "when next observed, the mould spores which had gained access had developed into a large colony...what was surprising was that the *staphylococcal* colonies in the neighbourhood of the mould, which had been well developed, were observed now to be showing signs of dissolution ..." The mould was identified as *Penicillium notatum* (the name was coined from the Latin word for paintbrush, due to the brush-like composition of the mould) and the active ingredient was called Penicillin.

While luck played a part in this observation of the lysis of the bacterial cells, Fleming used his experience and acumen to capitalise on his good fortune. However, the next 10 years showed unexpectedly slow progress towards the use of Penicillin in the treatment of bacterial infections. There are probably a number of reasons for this hiatus in developments. For a start, obtaining dilute aqueous solutions of penicillin from the mould proved difficult and nonreproducible. In addition, the *in vivo* stability of Penicillin was regarded as unsatisfactory. However, it was probably other scientists, some clinicians and certain medical doctors, suspicious of using a fungal broth to cure infections, who were mainly responsible for the loss of momentum. Nevertheless, some treatments were noted in this period. For example, a medical student at St Mary's contracted pneumococcal conjunctivitis which Fleming successfully treated with a Penicillin broth.

The outbreak of the Second World War revealed a need for a safe antibacterial substance for the treatment of infected deep wounds. A great deal of work was undertaken in the UK and the USA to prove that Penicillin was an ideal substance for such therapy. In Oxford, Florey and Chain et al continued to develop production, isolation and assay methods for Penicillin. They also demonstrated the potential of Penicillin by the successful treatment of experimentally infected animals.

In the North Regional Research Laboratory, Peoria, Illinois, strains of Penicillium which gave higher yields of Penicillin than Fleming's original culture were discovered (highly productive moulds were found on a piece of cheese and on an over-ripe melon). The American group showed that the mould could be grown in submerged culture with a consequent decrease in the number of other products formed. Better yields of Penicillin were obtained on adding corn steep liquor to the brew.

In the mid-1940s, the structure of Penicillin was elucidated. Significant contributions were made by Dorothy Hodgkin and Robert Robinson in Britain, and by the young American chemist R.B. Woodward. The bicyclic structure of a Penicillin contains a β-lactam ring that is readily cleaved; the endocyclic amide bond is the susceptible linkage. Simple β-lactams, first made by Staudinger in the early years of this century, are considerably more stable than Penicillin. The instability of the antibiotic was attributed to the steric strain imposed by the presence of the second sulphur-containing ring.

Since this period, other naturally occurring compounds possessing a β-lactam ring and exhibiting anti-bacterial properties have been isolated. In 1961, a strain of *Cephalosporium* growing near the outlet of a Sardinian sewer was shown to produce Cephalosporin C. For many years, it seemed that the Penicillins and cephalosporins were the only β-lactams to possess potent anti-bacterial properties. However, the isolation of Thienamycin, a very active and broad spectrum antibiotic, from Streptomyces and the discovery of the less active monocyclic compound Nocardicin from some strains of Nocardia dispelled this idea. The more interesting monocyclic compounds, called mono-bactams for example, Aztreonam have been described recently. The first members of the latter series of compounds were produced from bacteria found in soil samples taken from pine forests in the United States.

Preparation of semi-synthetic and synthetic β-lactams

Left to its own devices, the Penicillium fungus will produce isopenicillin N and Penicillin F. Supplying additional carboxylic acid to the fermentation broth gave modified penicillins. For example, addition of phenoxyacetic acid gave Penicillin V, while the presence of phenylacetic acid in corn steep liquor in the brew gave rise to Penicillin G.

Although of major importance in the clinic for a time, penicillins V and G became increasingly ineffective against a number of important pathogenic bacteria. The main reason for the developing immunity of some bacteria was their ability to produce an enzyme (a β-

lactamase) which was able to destroy the β-lactam before it could interact with and kill the bacterium. The production of penicillins with increased antibacterial activity was made possible by modification of the readily available natural products. First, scientists at the Beecham Laboratories found that the parent substance, 6-aminopenicillanic acid or 6-APA, could be obtained from some fermentations. Slightly later, 6-APA became available in bulk quantities by enzymic or chemical hydrolysis of the natural penicillins.

A vast range of semi-synthetic penicillins became available by chemical derivatisation of the free amino group. In this way, β-lactams were obtained that possessed broad spectra of activity against Gram-positive and Gram-negative organisms. The penicillins Amoxycillin and Ampicillin are of this type and still in widespread use. These new penicillins could be administered orally just like penicillins G and V, and they had a little extra stability against the β-lactamase enzymes. Other penicillins were prepared, with different side-chains at C-6, which conferred respectable antibacterial activity coupled with good β-lactamase stability on these molecules.

Another strategy has been developed recently to enhance the chance of success of Penicillin treatment against invasive, β-lactamase-producing bacterial organisms. This strategy involves administration of the Penicillin (for example, Amoxycillin) together with a second compound that inhibits the action of the β-lactamase enzyme. The β-lactamase inhibitor presently in use in the clinic is the β-lactam Clavulanic acid.

The chemistry of the penicillins and cephalosporins has been investigated in depth and a large number of intriguing transformations have been reported. Certain Penicillin derivatives can be transformed readily into the corresponding cephalosporin derivative lacking the acetoxy group linked through a carbon atom to the 3-position. Selective functionalization of the 3-methyl group is difficult and some attempts to accomplish straightforward transformations have led to unexpected rearrangements.

In short, chemical production of a wide range of cephalosporins from readily available penicillins is not attractive from a commercial standpoint. Furthermore, no simple enzymatic or fermentation process is available to modify the 7-amino side-chain of the cephalosporins in a similar fashion that described above for the penicillin system. Production of semi-synthetic cephalosporins has been accomplished by making cephalosporin C by fermentation, chemical removal of the acyl side chain to give 7-aminocephalosporanic acid or 7-ACA (usually with the carboxylic acid group protected as an ester) followed by acylation of the amino group and modification of the 3-substituent.

A wide variety of cephalosporins have been produced in this manner. Note that the 7-acylamino side-chains of the active cephalosporins differ in structure from the preferred 6-acylamino side-chains in the penicillin series. As few cephalosporins are active after oral administration, research in this field has been directed at finding broad spectrum agents for use by injection. Recently, the anti-bacterial agent Ceftazidime has been prepared. This cephalosporin is a powerful, broad spectrum, injectable antibiotic for use in the hospital environment and it is of particular value in the treatment of serious and life-threatening infections.

Words and Expressions

memorabilia	[复] n. 大事记,值得纪念的事	stopwatch	n. 秒表,跑表
tribute	n. 颂词,礼物,贡品	dedication	n. 贡献,奉献
fungal	adj. (=fungous)真菌的	mycobiological	adj. 真菌生物学的
spore	n. 孢子;vi. 长孢子	mould	n. 霉,霉菌
Staphylococcus/[复] *staphylococci*	n. 葡萄球菌	colony	n. (生物)群体
staphylococcal	adj. 葡萄球菌的	paintbrush	n. 画笔,漆刷
penicillium notatum	青霉菌,特异青霉,点青霉	ingredient	n. 成分,因素;(化合物的)组成部分
lysis/[复] lyses	n. (病的)渐退,消散,细胞溶解	acumen	n. 敏锐,聪明
nonreproducible	adj. 不能繁殖的,不能再生产的	hiatus	n. 脱落,裂缝
clinician	n. 临诊医师,门诊医师	momentum	n. 要素,动力
pneumococcal	adj. 肺炎双球菌的	melon	n.(各种的)瓜
conjunctivitis	n. 结膜炎	bicyclic	adj. 双环的
corn steep liquor	玉米浸液	endocyclic	adj. 桥环的
Nocardicin	诺卡杀菌素,诺卡地菌素	monocyclic	adj. 单环的
cephalosporium/[复]cephalosporia	n. 假头状孢子头	sulphur	n. 硫,硫黄
acetoxy group	乙酰氧基	Thienamycin	n. 甲砜霉素
Nocardia	n. 诺卡氏菌属,土壤细菌属	Streptomyces	n. 链霉菌素
mono-bactam	单内酰胺化合物	Aztreonam	n. 氨曲南
Ceftazidime	n. 头孢他啶	semi-synthetic	半合成的
carboxylic acid	羧酸	isopenicillin	n. 异青霉素
phenoxyacetic acid	苯氧基乙酸	Amoxycillin	n. 阿莫西林
phenylacetic acid	苯基乙酸	Ampicillin	n. 氨苄西林
6-aminopenicillanic acid	6-氨基青霉烷酸	acyl	n. 酰基
7-aminocephalosporanic acid	7-氨基头孢烷酸	depict	vt. 描述,描写

invasive	adj. 侵略的,侵害的,侵袭的	rearrangement	n. 重排
Clavulanic acid	克拉维酸	acylamino	n. 乙酰氨基
transformation	n. 变化,转化,转换,蜕变	side-chain	侧链

straightforward　*adj.*;*adv.* 简单的(地),易懂的(地),坦率的(地)
sardinian　*n.* 撒丁岛人,撒丁岛语;撒丁岛的,撒丁岛人的
intriguing　*adj.* 有趣的,有迷惑力的;*v.* 激起……的兴趣
fungus　*n.* 真菌(包括霉菌、酵母菌和伞菌等)

Notes

① The name was coined from the Latin word for paintbrush, due to the brush-like composition of the mould. 根据英汉两种语言存在一词多类、一词多义现象,在翻译过程中,需要通过上下文、词在句中的搭配关系以及词在句中的词类来选择恰当词义,使译文准确无误。如本句中 composition 有"构造、组成"等词义,根据其前面的单词 brush-like(意思是像刷子一样),可以确定 composition 应该选择"构造"词义。

② However, it was probably other scientists, some clinicians and certain medical doctors, suspicious of using a fungal broth to cure infections, who were mainly responsible for the loss of momentum. 此句比较长,分析可知这是"it is...who..."的典型强调句。前半句中,some clinicians and certain medical doctors 与 other scientists 是同位语,目的是对 other scientists 进行进一步说明是哪些其他科学家,而其后面的形容词短语 suspicious of using a fungal broth to cure infections 则用来解释其他科学家做了什么(怀疑用真菌肉汤治疗感染)。

③ β-lactams were obtained that possessed broad spectra of activity against Gram positive and Gram negative organisms. 主句为被动句,翻译时采用主谓变序法,译成无主语句,也就是将英语句中的主语译成汉语中的宾语,即先译谓语,后译主语(obtained β-lactams)。此外,posess...activity 表示"具有……活性"的意思;broad spectra 表示"广谱",broad spectra of activity 意思是广谱活性。

④ The *in vivo* stability of Penicillin was regarded as unsatisfactory.
青霉素的体内稳定性不理想。

Exercises

Answer the following questions:

a. What affected the application progress of penicillin in the treatment of bacterial infections after Fleming discovered penicillin?

b. Does people satisfy the stability of antibiotics? Why?

c. Why did the efficacy of penicillins V and G get worse against some pathogenic bacteria?

Reading material A:

Origins of the Quinolone Class of Antibacterials: An Expanded "Discovery Story"

The following excerpt from a 2005 review of the field of antibacterial quinolones is representative of descriptions in the scientific literature concerning the origin of the class:

"The first antimicrobial quinolone was discovered about 50 years ago as an impurity in the chemical manufacture of a batch of the antimalarial agent chloroquine (Figure 1-4). It demonstrated anti Gram-negative antibacterial activity, but its potency and antimicrobial spectrum were not significant enough to be useful in therapy. Building on this lead, however, subsequently nalidixic acid was commercialized."

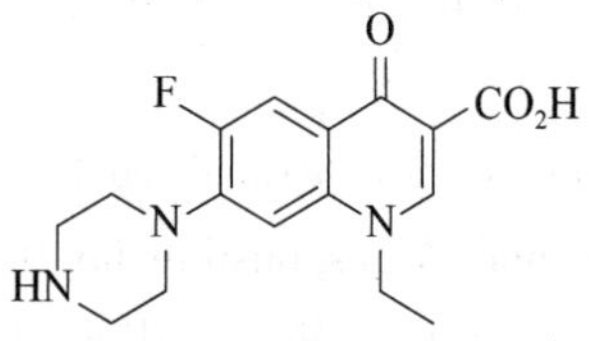

3
norfloxacin

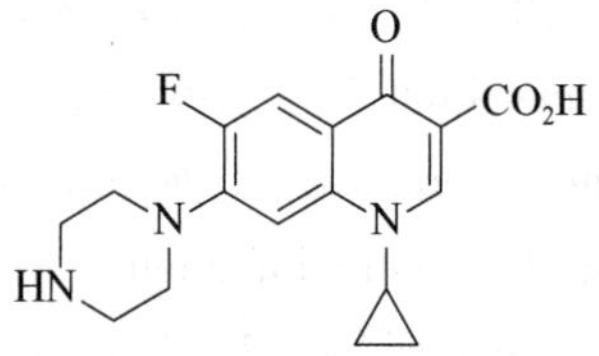

4
ciprofloxacin

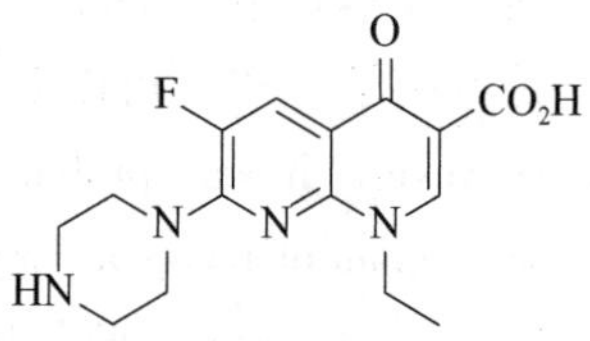

5
enoxacin

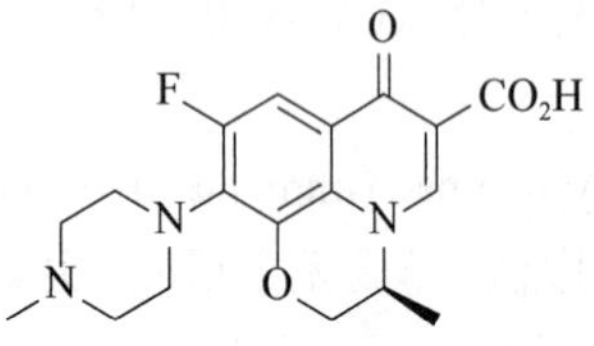

6
levofloxacin

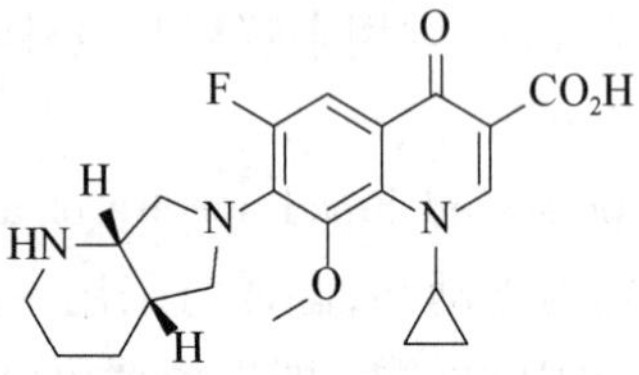

7
moxifloxacin

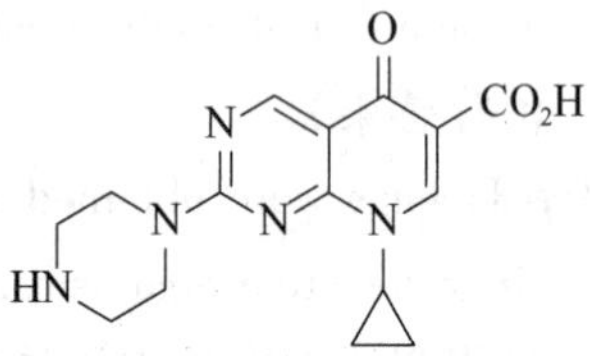

8
pipemidic acid

Figure 1-4 Structures of representative quinolones (including core variants) introduced into clinical practice over the past several decades

"Figure 1-4" from that review depicts the structure of the key impurity 1, a quinolone core compound, as well as the structure of nalidixic acid 2, a 1, 8-naphthyridone core compound (see Figure 1-5 in the current review). As explained in more detail below, the above excerpt surprisingly encompasses essentially all the information concerning the origin of nalidixic acid published by its discoverers at Sterling Drug (now part of Sanofi) and moreover may even be making inferences beyond the data available from Sterling's published literature (i.e., "... potency and antimicrobial spectrum [of compound 1] were not significant enough to be useful in therapy"). For many purposes, as for inclusion in review articles and general scientific books on pharmaceuticals and anti-infectives, summaries that convey the "discovery story" in one or two sentences may be adequate. However, for scientists involved in the strategy of drug discovery wishing to know the underlying rationale and data leading to the selection of significant first-inclass drugs (in this case nalidixic acid) they are inadequate, as there are many highly pertinent questions that remain unanswered. For nalidixic acid, such questions include the following: (1) What SAR or other influences led to the switch from the quinolone core of lead compound ("impurity") 1 to the 1, 8-naphthyridone core of the launched drug nalidixic acid? (2) What is the antibacterial potency and spectrum of 1, and in what ways did nalidixic acid improve upon it? (3) What is the chemical mechanism for the formation of 1 as an impurity? Additionally, from a general context and priority point of view, the following questions require clear answers as well: (4) Was the 3-carboxy substituted quinolone or naphthyridone core a unique chemical structure at the time of Sterling's 1962 disclosure of nalidixic, and if not, had antibacterial activity previously been reported with similar structures? (5) If antibacterial activity had previously been observed with such structures, what led to that independent discovery, why was that effort largely overlooked in the quinolone scientific literature, and why did launched drugs not emerge directly from those efforts? On the basis of primary literature from a number of disparate sources, this review will provide answers to most of these key questions. In some instances definitive answers are still not possible, but the author has made educated guesses pertaining to the drug discovery logic that might have been applied at that time which then resulted in certain events and decisions. Those instances wherein informed speculation must substitute for available facts will be clearly highlighted as such.

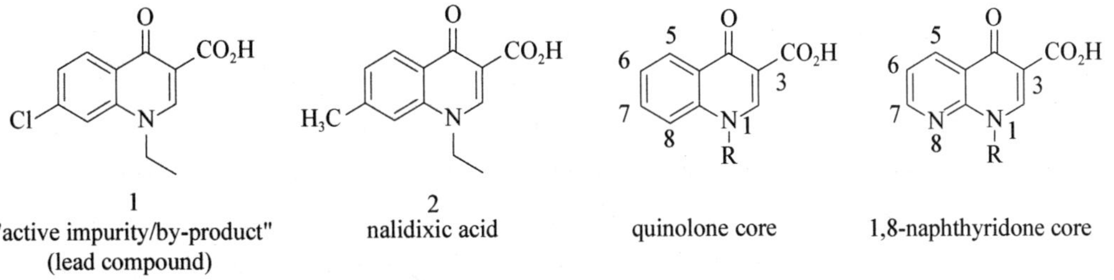

Structure of the chloroquine synthesis "active impurity" ("byproduct") 1 which led Sterling Drug to the identification of nalidixic acid 2; quinolone 1 was also listed as a patent example in ICI's GB830832.

Figure 1-5 The core structures and numbering of quinolones and corresponding 1, 8-naphthyridones

For the purpose of wider context, the following is a brief account of the key achievements over 5 decades in the antibacterial quinolone field after the launch of nalidixic acid. For further information on this subject the reader is directed to key reviews and books selected from a vast literature. Both quinolone and 1, 8-naphthyridone based antibacterial drugs today are often informally included within the broad, generally interchangeable designations "quinolone" or "fluoroquinolone" antibacterial class; occasionally, the 1, 8-naphthyridone core is referred to as an 8-azaquinolone. In the 15 or so years following the 1964 clinical introduction in the United States of nalidixic acid by Sterling Drug, a number of follow-on agents were launched by other companies (Sterling itself later launched two additional drugs from this class for human use, rosoxacin and amifloxacin). At first, agents within this class occupied a fairly narrow therapeutic niche, being used primarily for treatment of urinary tract infections caused by *Escherichia coli* and a few other Gram negative pathogens. However, the therapeutic utility of the class increased dramatically starting in the early 1980s following the discovery that substitution at the quinolone (or 1, 8-naphthyridone) 6-position with fluorine and at the 7-position with a basic amino heterocyclic group together greatly enhanced the antimicrobial potency and expanded the microbiological spectrum of these agents. These fluoroquinolones allowed effective treatment of infections caused by significant Gram negative pathogens such as *Pseudomonas aeruginosa*. Norfloxacin (3) was the first such fluoroquinolone (Figure 1-4). Over the following decades, pharmaceutical companies have launched agents derived from both core types (e.g., ciprofloxacin 4, a quinolone, and enoxacin 5, a 1, 8-naphthyridone, Figure 1-4). There appears to be no firm consensus in the field to suggest whether the naphthyridone compared to the quinolone core offers any significant intrinsic advantages for therapy. Rather, choice of peripheral substituents dominates effects on antibacterial potency, spectrum, pharmacokinetics, and safety more than the choice of quinolone vs naphthyridone core. The quinolone core does, however, have the technical advantage of the availability of the 8-carbon as an additional site for substitution, a feature exploited in several commercially successful quinolones, among them levofloxacin 6 and moxifloxacin 7. Although many other core variations besides quinolones and 1, 8-naphthyridones have been tested, and several such variations have even been launched (e.g., pipemidic acid 8), the quinolone and naphthyridone-based agents have remained the dominant core variations within the class. During the past 2 decades or so, launches of new agents having the quinolone core have surpassed launches of new 1, 8-naphthyridone-based drugs Over the years, the clinically useful microbiological spectrum of the quinolone class has expanded further to include many Gram positive pathogens, such as Staphylococcus aureus. Although marketed quinolones have proven to have a favorable safety profile, adverse events have inevitably arisen within this class of drugs, as with most classes of drugs. As a result, a number of entrants have been withdrawn over the decades or their use restricted because of various reasons (cardiovascular issues or hepatotoxicity for example). Moreover, therapy with quinolones is associated with an increased risk for tendinitis and tendon rupture. Nevertheless, within the field of anti-infectives, the number of clinical introductions within this class over the past 5 decades has been rivaled only by the number of introductions of all β-lactam antibacterials (penicillins, cephalosporins, and carbapenems). The quinolone class of antibacterials has been spectacularly successful from both a medical and a commercial point of view. Therefore, the first-in-class introduction of nalidixic acid by Sterling must be regarded as a highly important pharmaceutical achievement, having a remarkable and long-lasting positive influence on medicine.

引自：BISACCHI G S. Origins of the quinolone class of antibacterials: an expanded "discovery story"[J]. Journal of Medicinal Chemistry, 2015, 58(12): 4874-4882.

Words and Expressions

antibacterial	*adj.* 抗菌的	quinolone	*n.* 喹诺酮类
antimicrobial	*adj.* 抗菌的	manufacture	*v.* 制造
carboxy	*n.* 羰基	spectrum	*n.* 谱，范围
impurity	*n.* 杂质	therapy	*n.* 治疗
logic	*n.* 逻辑	naphthyridone	*n.* 萘啶酮
pharmaceutical	*adj.* 制药的，药学的；*n.* 药品，药剂	urinary	*adj.* 泌尿的
pathogens	*n.* 病原体	interchangeable	*adj.* 可交换的
levofloxacin	*n.* 左氧氟沙星	fluorine	*n.* 氟
tendon	*n.* 肌腱	Norfloxacin	*n.* 诺氟沙星
Moxifloxacin	*n.* 莫西沙星	heterocyclic	*adj.* 杂环的
Ciprofloxacin	*n.* 环丙沙星	infective	*adj.* 感染的
Enoxacin	*n.* 依诺沙星	pipemidic acid	吡哌酸
Escherichia coli	大肠杆菌	nalidixic acid	萘啶酸
staphylococcus aureus	金黄色葡萄球菌	*pseudomonas aeruginosa*	铜绿假单胞菌

Notes

① It demonstrated anti Gram-negative antibacterial activity, but its potency and antimicrobial spectrum were not significant enough to be useful in therapy.

它显示出抗革兰氏阴性菌活性，但其效力和抗菌谱不足以用于治疗。

② SAR

构效关系

③ Therefore, the first-in-class introduction of nalidixic acid by Sterling must be regarded as a highly important pharmaceutical achievement, having a remarkable and long-lasting positive influence on medicine.

因此，Sterling 的首创新药萘啶酸被视为一项极其重要的药学成就，对医药产生显著且持久的积极影响。

first in class 为医药中的专用术语，针对某个靶点和适应证的首创用药，即能治疗某个疾病的第一种新药，属于一类原研新药，具有里程碑意义。

Exercises

1. Put the following into English:

抗菌　　杂质　　感染　　制药的　　病原体

2. Put the following into Chinese:

a. The discovery and introduction into clinical use of entirely new and useful classes of medicine arguably represent the highest pinnacle of achievement for pharmaceutical researchers in the field.

b. The publication of ICI's GB830832 in March 1960 could have then motivated Sterling to initiate a traditional "fast-follow" medicinal chemistry program by first developing SAR around the quinolone-core series to provide benchmark data, followed by extending that SAR to patent-unencumbered 1, 8-naphthyridone compounds.

c. Over the following years, however, data became available in the literature that can now be used to effectively provide "head-to-head" minimum inhibitory concentration (MIC) comparisons of nalidixic acid 2 versus the chloroquine synthesis byproduct quinolone 1 as well as versus the ICI "matched pair" nalidixic acid quinolone analog 12.

Reading material B:

Chinese Treatments Benefit World

Unit 3

Chemotherapy: An Introduction (I)

Chemotherapy can be defined as the use of chemical compounds to destroy infective parasites or organisms without destroying their animal host. Ancient literature describes the chemotherapeutic preparations of early times, but many of these were worthless medicines associated with superstitions and magic. Some of these compounds, however, were shown to have value through the process of trial and error over many years. In 3000 B.C., the Chinese emperor Shen Nong noted many curative substances in the *Book of Herbs*. "Chang Shan" was stated to be of value against malarial paroxysms and related fevers. Since that early time, its antimalarial activity has been confirmed and its efficacy corroborated by present-day investigators.

Germ theory of disease

In the nineteenth century, the germ theory of disease became established. Agostino Bassi of Lodi demonstrated the transmission of silk-worm disease by a pathogenic microorganism, and proposed the transmission of certain human diseases by this mechanism. Davaine concluded that anthrax in animals was caused by bacteria, and this was later proved by Pasteur. In 1865, Lister demonstrated the sterilization properties of phenol, and the medical profession started to accept the germ theory of disease. A new era in medicine began.

The influence of Paul Ehrlich on the development of chemotherapy first appeared at the turn of the century. As a result of his discoveries, he is regarded as the father of chemotherapy. He first examined the distribution of dye materials in the blood and subsequently in living animals. Dyes were chosen because they were easy to see after distribution in the animal. Ehrlich found that certain dyes colored certain organs or systems selectively, whereas other dyes stained tissues generally. As the science of bacteriology developed, Ehrlich turned his attention to the staining of bacteria. He developed the acid-fast stain for the tubercle bacillus, and performed much of the early work leading to the use of the Gram stain.

Phenols

Lister applied Pasteur's germ theory to surgery in 1865. He showed that cleanliness and sterilization of wounds with phenol could prevent the dreaded putrefaction, which was common in those days. This provided the incentive for further investigations in this field. It was subsequently shown that phenols that had been alkylated or halogenated had increased antibacterial potency. In 1932, research on the bis-phenolic compounds began, and this resulted in the patenting of hexachlorophene in 1941.

Dyes and arsenicals

In 1887, Rozahegyi reported that certain strains of bacteria did not grow on nutrient agar in the presence of certain dyes. In 1890, Stilling reported aniline dyes to be highly active antibacterial substances. In 1891, Ehrlich found that methylene blue stained malarial organisms, but only limited success was obtained when the dye was tried on malarial patients.

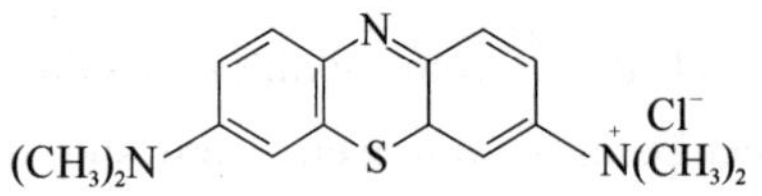

Methy lene blue

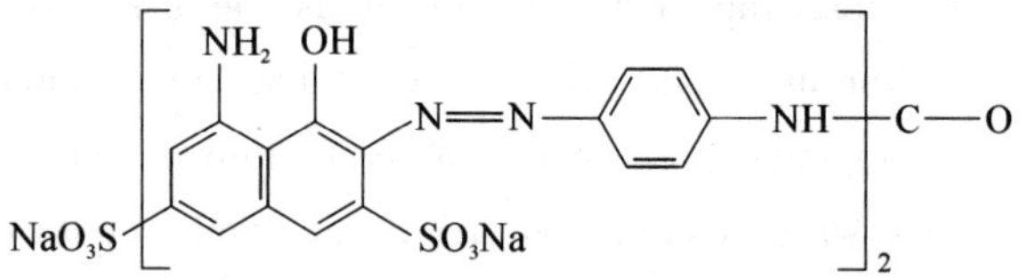

Afridol violet

Because sleeping sickness was a major problem in the development of Africa by Europeans, Ehrlich attempted to find a curative dye for trypanosomiasis. Trypan red, an azonaphthalenesulfonic acid derivative, was found effective against certain species of trypanosomes, but had limited effectiveness against other species. Other dyes of this series, such as trypan blue and afridol violet, however, were later found to be of more practical value.

Resistant strains of microorganisms later developed. Resistance to drugs containing arsenic was also observed. The parasites that were resistant to one class of compounds, however, were not resistant to another class. Ehrlich explained this by his chemoreceptor theory. If the receptor of the parasite had reduced affinity for one class of drugs, it could still combine with another class. This also suggested the possibility of different types of receptors.

In 1859, Bechamp heated aniline and arsenic trioxide and obtained a compound believed to be the anilide. This is represented by equation (1).

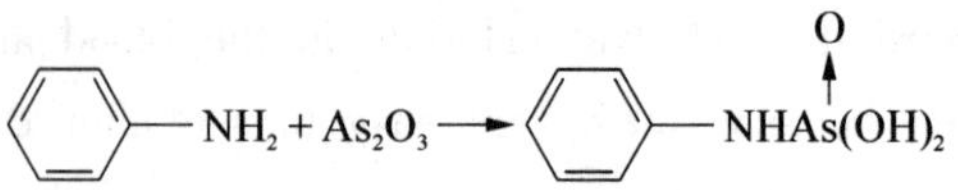

(1)

In 1903, Ehrlich tested this compound and assumed the previous structure to be correct. He found that it was inactive *in vitro* against trypanosomes, but it was not tested *in vivo*. In 1905, it was tested against trypanosomes *in vivo* by Thomas and Breinl, who found that it was not only active but 40 times less toxic than potassium arsenite. They named the compound atoxyl. It was shown to be effective against sleeping sickness organisms by Koch, who was then working in Africa. Because of the demonstrated effectiveness *in vivo*, Ehrlich's interest in this arsenical compound was revived. He then showed the correct structure of atoxyl. In 1909, Ehrlich showed that atoxyl, when reduced to the trivalent arsenous state, did have good trypanosomicidal activity *in vitro*. He then proposed that the host cells reduced the pentavalent arsenic to the trivalent state, and that this was the active form of atoxyl.

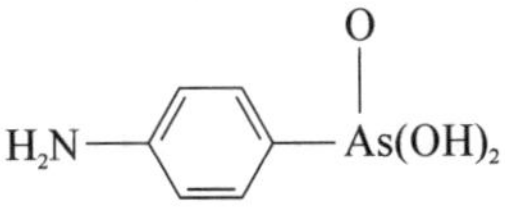

Atosyl 1

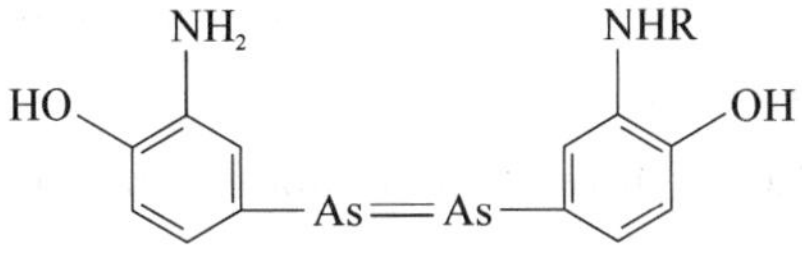

Salvarsan,R=H
Neosalvarsan,R=CH_2SO_2Na

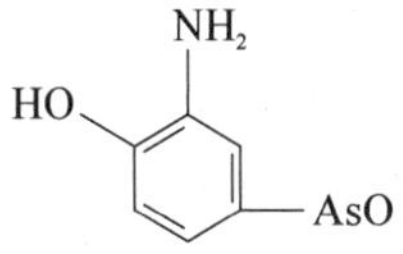

Oxophenarsin

In 1910, Ehrlich introduced a trivalent arsenical for the treatment of syphilis; this compound, known as arsphenamine or Salvarsan, was a major triumph of chemotherapy. A less toxic derivative was Neosalvarsan; neither of these compounds could be obtained in a pure state. The corresponding arsenoxide, oxophenarsine, became more widely used.

The activity of the arsenical drugs is explained as due to a blocking of essential thiol groups. For example, lipoic acid dehydrogenase contains two cysteine molecules, which are kept near each other by folding of the molecule. As a result, an arsenical can react with these thiol groups and inactivate the molecule, as shown in equation (2).

$$RAs(OH)_2 + 2HSR \longrightarrow RAs(SR)_2 + 2H_2O \qquad (2)$$

Other heavy metal compounds

In addition to the arsenicals, other compounds containing heavy metals are active chemotherapeutic agents. It is considered that the bismuth compounds act against parasites as the arsenicals do, by reaction with cellular thiols.

Antimony potassium tartrate, also known as tartar emetic, was shown to be effective against leishmaniasis in 1908. Soon the treatment of this condition with this antimonial was common, but because of the high toxicity of the trivalent antimony potassium tartrate, the pentavalent antimony compounds replaced it. The arylstibonic acids were among the first pentavalent antimonials to be used.

The antimony ions are believed to combine with the mercapto groups of the microorganism's phosphofructokinase. In the host, this enzyme is apparently sufficiently different that antimony does not react to the same degree. This inhibition of phosphofructokinase results in accumulation of fructose-6-phosphate. As a result, a major source of energy is denied the infecting organism.

Antimalarials

The history of Cinchona bark as an antimalarial drug is connected with many folk tales.

A well-known story involves the discovery of the antimalarial effect of cinchona bark. A Peruvian Indian who was stricken with the fever drank from a stagnant pond into which several trees had fallen. Apparently, the alkaloids from these trees had dissolved in the water. Within hours, the Indian's fever subsided, and he eventually recovered. The news of this cure spread, and the bark of these trees was used by the natives. The Jesuit missionaries learned of the use of this bark to treat fever from malaria.

The Countess Ana de Osorio, wife of the Count of Chinchon and Viceroy of Peru, was a victim of tertian fever, and used this bark successfully. It was introduced in Europe in 1633, and its use was further spread by the Jesuit order. Early names for the bark were Countess' bark, Jesuit's bark, and Peruvian bark. Linnaeus wished to honor the Countess of Chinchon when he named the bark, but he omitted the second letter in the name. As a result, it is called cinchona bark today.

It has been reported that the Countess of Chinchon died before she got to Peru. The second wife of the Count of Chinchon never had malaria while she lived in Peru. The Count contracted the fever and was treated by bleeding, which was the customary treatment of the time; he was not cured by any drug. The story therefore remains open to question.

The alkaloid quinine was isolated from cinchona bark in 1820, and was used in the treatment of malaria until synthetic drugs were developed.

Researchers in Germany became interested in the treatment of malaria early in the century. Following the limited success reported by Ehrlich in the treatment of malaria with methylene blue, alterations of the methylene blue molecule by German workers produced no compounds of practical value, but the experience they gained was applied to other ring systems. As a result, pamaquine was reported in 1924 to be an effective anti-malarial drug. Pamaquine was first accepted with enthusiasm, but was later found to be too toxic and inferior to Quinine. It was no longer used by 1930.

Quinacrine (Atabrine) was prepared by Mauss and Mietzsch in 1933. Its use became widespread during the war years of 1939 to 1945 because the Japanese controlled most of the Quinine supply. Quinacrine was found to be an effective antimalarial with low toxicity. Between 1941 and 1945, university and industrial laboratories cooperated in an antimalarial program in the United States. Nearly 13000 new organic compounds were synthesized, analyzed, and tested, and some were clinically evaluated during this period. Each compound was given a survey number, abbreviated SN. For example, Quinacrine was also known as SN 390.

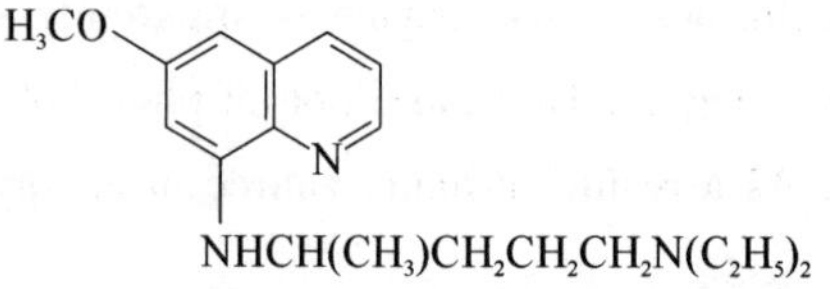

Pamaquine

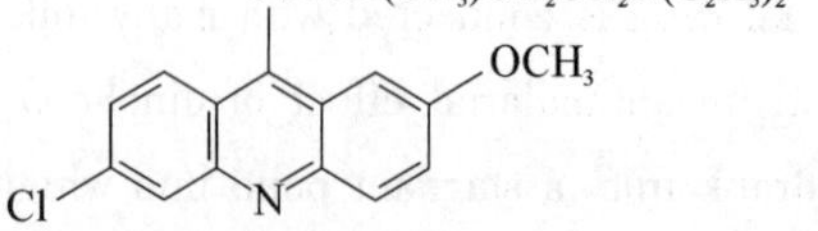

Quinacrine

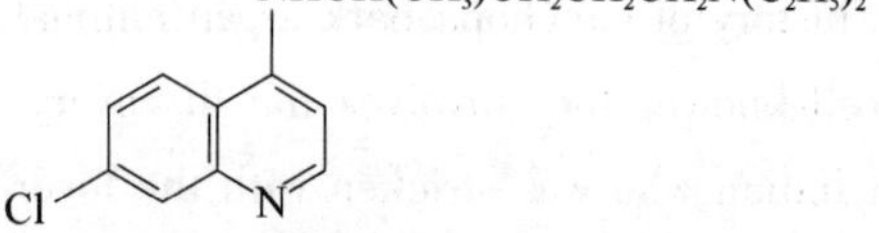

Chloroquine

Earlier, in 1939, Chloroquine was prepared by chemists at the Bayer compound in Germany. A limited amount was sent to Tunis, which was under German domination at that time. After the Anglo-American forces took control of this area in 1943, a sample of Chloroquine was turned over to them. It was found superior to Quinacrine for the treatment of malaria.

Schoenhofer postulated in 1942 that the possibility of tautomerism was necessary for antimalarial activity. A series of pyrimidine derivatives was prepared as potential antimalarial drugs, and their activity was attributed to their tautomeric possibilities. When an amino-substituted pyrimidine ring is opened, a biguanide results, which also has the tautomeric forms regarded as desirable. High activity was found for this type of compound, which prompted the investigation of other biguanides. This led to the discovery of valuable compounds, including Chlorguanide, which was introduced as an antimalarial in 1946.

Cyclization of this biguanide produced Cycloguanil, which was also found to be a highly active antimalarial.

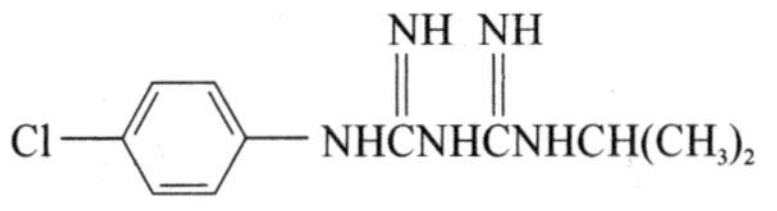

Chlorguanide

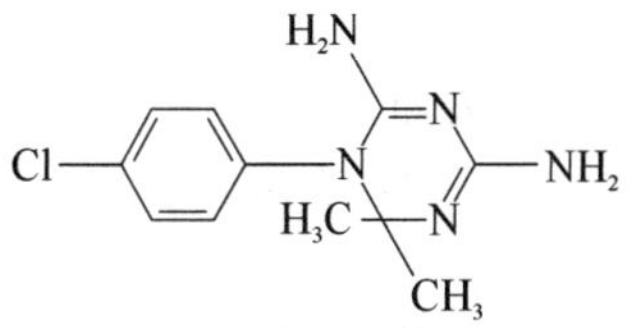

Cycloguanil

Resistant strains of malarial parasites found in Vietnam have caused renewed interest in malarial chemotherapy, and new classes of antimalarials are being investigated.

Trypanosomicides

Trypan red was formerly used as a trypanosomicidal agent in the treatment of African sleeping sickness. Use of a dye was found objectionable, however, and the colorless suramin sodium replaced it.

Trypanosomes have a high rate of carbohydrate metabolism, and it was concluded that suramin sodium was trypanosomicidal because of its antagonism of carbohydrate metabolism. In 1926, it was reported that some Guanidine derivatives lowered blood sugar levels in animals. These compounds were successful as trypanosomicidal agents, but they acted at extreme dilutions, and insulin was found to be ineffective against Trypanosomes. Because of this, Lourie and Yorke believed that a Guanidine acted directly on the Trypanosomes, and its activity was independent of its hypoglycemic action.

$$R—(CH_2)_n—R,\ R—C(=NH)—NH_2,\ —S—C(=NH)—NH_2,\ —NH—C(=NH)—NH_2,$$

$$H_2N—C(=NH)—NH(CH_2)_{10}NH—C(=NH)—NH_2$$

Synthalin

Large numbers of guanidines, amidines, amines, and isothioureas were then investigated as trypanosomicidal agents, the most important being Synthalin. Certain diamidines in this series, particularly Synthalin, showed a high activity both *in vitro* and *in vivo*. Because these agents were unable to penetrate to the. central nervous system (CNS), they were found of little value in the treatment of late stages of sleeping sickness.

Words and Expressions

paroxysm	*n.* (病)发作,突发,暴发	chemotherapy	*n.* 化学治疗法,化疗
bacillus	*n.* 杆状细菌;杆菌	superstition	*n.* 迷信
putrefaction	*n.* 腐烂,腐败,腐败物	nutrient	*adj.* ; *n.* 营养的(物),滋养的(物)
pathogenic	*adj.* 致病的,病源的	agar	*n.* 洋菜,石花菜,细菌培养基
anthrax	*n.* 炭疽	affinity	*n.* 亲和力,嗜好
sterilization	*n.* 消毒,杀菌	antagonism	*n.* 对抗(性,作用),对立性
arsenic	*n.* 砷	Quinacrine	*n.* 米帕林
tubercle	*n.* 结核	tertian	*adj.* 间日的,隔日(发作)
syphilis	*n.* 梅毒	*in vitro*	在体外,在玻璃试管内,在玻璃容器内
in vivo	在体内,自然条件下的(实验,化验)	tautomerism	*n.* 同质异构,互变(异构)现象,互变异构
thiol	*n.* 硫醇类	Tubercle bacillus	结核菌

Notes

① the Chinese emperor Shen Nong: 翻译为炎帝神农,是中国神话传说中司农业、医药的神。在此处,不能把 Chinese emperor 翻译为皇帝。

② The influence of Paul Ehrlich on the development of chemotherapy first appeared at the turn of the century. 首先,the influence of A on B 是制药工程、化学等专业常用表达方式,用来表示某些因素 A 对 B 的影响,比如 the influence of temperature on the yield of product 温度对产物产率的影响。其次,本句中 at the turn of the century 翻译时不能仅翻译为"在世纪之交",需要具体指出是哪个世纪之际,或者哪个世纪末/初。

③ Ehrlich found that certain dyes colored certain organs or systems selectively, whereas other dyes stained tissues generally. 这是一个既简单又容易被学生翻译错的句子,关键点是副词的翻译。副词主要用来修饰形容词、动词,本句中二个副词 selectively 和 generally 分别修饰同义动词 color 和 stain。

④ Soon the treatment of this condition with this antimonial was common. 英语中常见一种现象是使用具有动作意义的名称表达动词的含义。由于中、英语言习惯差异，将这类名称翻译成中文的时候习惯将其翻译为动词，也就是翻译技巧中的词类转换法。所以本句中 treatment 应该翻译为动词治疗。

⑤ Tautomerism: 互变异构现象，指存在一种物质，它是两种可以相互转化形态的平衡混合物，这种转化一般是由于某个氢原子的迁移，因此互变异构化合物可以产生两个系列的衍生物。

Exercises

1. Put the following into English:

衍生物	硫醇	互变异构体	波谱
染料	微生物	苯胺	消毒

2. Put the following into Chinese:

malarial	chemotherapy	bacteria	*in vitro*
antagonism	affinity	parasite	putrefaction

Reading material A:

Chemotherapy: An Introduction (Ⅱ)

Sulfonamides

In 1935, Prontosil was introduced as a synthetic antibacterial agent, and a new era of bacterial chemotherapy was opened. Prontosil was inactive *in vitro* but did have good activity *in vivo*, particularly against hemolytic streptococcal infections.

The history of Prontosil goes back to earlier work on azo dyes. In 1909, it was noted that dyes containing a sulfonamide group formed stable complexes with wool proteins. In 1919, Heidelberger and Jacobs tried to increase the antibacterial properties of hydrocupreine by coupling this molecule with sulfanilamide by an azo linkage. Some activity against the pneumonia organism was seen.

Shortly after the introduction of Prontosil, Trefouel and co-workers suggested that this compound was cleaved at the azo linkage in the host tissue and yielded sulfanilamide. Further work on the bacteriostatic effect of sulfanilamide showed that it was highly active by itself. In the following years, other sulfanilamide derivatives were prepared. In 1938, Sulfapyridine was shown to be even more effective than sulfanilamide. Research in this area continues still, and new sulfonamide drugs have been introduced.

Antibiotics

The term antibiotic was introduced by Waksman in 1942. An antibiotic can be defined as a chemical substance produced by microorganisms that can inhibit growth of, or even destroy, other microorganisms. It has been reported in folk literature that the Chinese treated infections such as boils and carbuncles with an extract of a mold curd made from soybeans. Moldy cheese has also been used historically by Chinese and Ukrainian peasants to treat infected wounds. Antibiotic agents were not studied systematically, however, until the twentieth century.

Pasteur and Joubert noted that anthrax bacilli were killed if certain common bacteria were grown with them. Injection of a deadly dose of anthrax bacilli into a laboratory animal was rendered harmless if common bacteria were injected at the same time. In 1890, the antibacterial extract of *Pseudomonas aeruginosa* was found of value in the treatment of diphtheria and other pyrogenic coccal infections. This was the pyocyanase of Emmerich. This product contained two antibiotic substances called pyocyanase and pyocyanine. This mixture was not satisfactory, however, and its use was discontinued shortly after the turn of the century.

Interest in antibiotic substances was revived in the 1930s, after the discovery by Fleming in 1929 that a filtrate of a broth culture of a penicillium mold had distinct antibacterial properties. A culture of staphylococcus organisms was accidentally contaminated with the spores of penicillium notatum. Around the colonies of the mold, well-developed growth of staphylococcus appeared to be dissolving. When the mold was isolated in a pure culture, it produced a material with a powerful *in vitro* effect against many common bacteria that caused infections in humans. It had no effect against some other bacteria.

Because the mold contaminant was known as penicillium notatum, the material it produced was called Penicillin. It had an extremely low toxicity, and Fleming suggested its use as an antiseptic. Systemic use of penicillin was not attempted at this time. Subsequent attempts by other workers to concentrate this material were not successful because of the instability and low concentrations of Penicillin in the filtrates of the broth. Consequently, Penicillin remained a scientific curiosity for about 10 years.

In 1938, Florey and Chain made a systematic survey of antibiotic substances. Because of the interesting chemical and biologic properties of Penicillin, it was selected as one of the first to be studied. These workers succeeded in purifying the crude substance and demonstrated its remarkable antibiotic properties in mice and humans with *staphylococcal* and other infections caused by gram-positive organisms.

During World War Ⅱ, England was under heavy air attack by the Germans, and this limited the development of penicillin production there. In 1941, Florey and Heatley went to the United States for experimental assistance in the production of Penicillin.

American scientists, however, had also read of Fleming's work in 1929 and had also done some research in this area. The Northern Regional Research Laboratories had even patented a procedure for the submerged fermentation production of Penicillin in 1935, six years before the visit of Florey and Heatley.

It was soon found that the Penicillin made in England and the United States were not identical. Phenylacetic acid could be obtained as a hydrolysis product of the Penicillin made in the United States but not from that made in England. Conversely, a hexenoic acid was obtained from the hydrolysis products of the British Penicillin but not from the American Penicillin. The British Penicillin became known as Penicillin Ⅰ (later known as Penicillin F) and the American product as Penicillin Ⅱ (later known as Penicillin G).

By 1943, only Penicillin G could be obtained in a pure state. Penicillin F contained small amounts of impurities and Penicillin G was more active. As a result, Penicillin G was produced in the United States after 1943.

After the introduction of Penicillin G, other less soluble salts were prepared to prolong the action of a single dose. Other new Penicillin derivatives that were more effective orally, or derivatives that resisted penicillinase, were also developed. In addition, Penicillin derivatives with a broader spectrum of antibacterial activity were produced. The cephalosporin relatives of these antibiotics have also been introduced in antibacterial therapy.

Searching the Actinomycetales order of microorganisms, Waksman found a valuable antibiotic in 1943. It was produced by a Streptomyces genus and was called Streptomycin. This drug has been particularly effective against mycobacteria. Neomycin was discovered in extracts of *Streptomyces fradiae* by Waksman and Lechevalier later in 1949.

In 1945, an antibiotic mixture of polypeptide substances was isolated from the infected tissues of a small girl named Margaret Tracy. This material, produced by *Bacillus subtilis*, was named Bacitracin

Chloramphenicol, a broad-spectrum antibiotic produced by streptomycetes, was found in a Venezuelan soil sample. It was isolated from fermented media in 1947, and later was produced synthetically on a commercial basis. It has been valuable in treating typhus and typhoid fever.

The first of the broad-spectrum tetracycline family of antibiotics to be discovered was Chlortetracycline, isolated by Duggar, a retired professor, in 1947. In 1950, Oxytetracycline was isolated, and the structural resemblance to Chlortetracycline was noted. After the structures were determined, a reductive dehalogenation of Chlortetracycline was performed in 1953, producing Tetracycline. This family of antibiotics has a wider range of activity than any of those previously discovered.

Anticancer agents

The preparation of mustard gas was described by Meyer in 1886, and its vesicant properties were noted. During World War Ⅰ, mustard gas was used by the military forces, and the vesicant action on the eyes, skin and respiratory tract was described. Autopsies of soldiers killed by the gas attacks showed toxic effects on the leukocytes, bone marrow, lymph tissue, and mucosa of the gastrointestinal tract. After World War Ⅰ, further research was carried on, and the closely related nitrogen mustards were developed. The activity of this type of compound in combating certain types of cancers was observed. At the close of World War Ⅱ, this information was declassified, and the anticancer potential of these compounds was made public. Soon, many variations of these alkylating agents were prepared as anticancer drugs. These variations included the ethylenimines, the alkylsulfonates, and later the nitrosoureas. Many of the modern anticancer agents have been developed from these agents.

Cancer cells, like normal cells, need specific compounds for their metabolism. It was felt that certain chemicals might resemble a normally occurring, essential compound but not be able to function exactly as the normal compound should. If this false chemical were incorporated into a metabolic process, the resultant product could be inactive. A cancer cell that incorporated such a false substance would thus be inhibited or killed. Based on this assumption, many antimetabolites have been prepared as anticancer agents, e.g., folic acid and methotrexate. Folic acid is needed in the normal metabolism of a cell, and methotrexate is an analog of folic acid. It closely resembles folic acid but blocks its normal function. Methotrexate was shown to be effective against leukemias in 1948.

Purines and pyrimidines are essential components of nucleic acids. If analogs of these compounds were prepared, they might also be falsely incorporated into the nucleic acids and thus block the normal function of the cell. Purine analogs were studied by Hitchings in 1942 as potential anticancer agents. From this series, mercaptopurine was described as a clinically effective anticancer agent in 1952. Pyrimidine analogs were also described as effective anticancer drugs in 1957; 5-Fluorouracil became a clinically effective drug.

A number of antibiotics have shown some activity against various types of cancers. The actinomycins have been the most clinically effective; they are powerful bacteriostatic agents exhibiting cytostatic properties. The first of these antibiotics, Actinomycin D, was isolated from actinomycetes in 1940.

Colchicine, from *Colchicum autumnale*, and Podophyllotoxin, from *Podophylum peltatum*, have long been known as inhibitors of cellular mitosis in the metaphase. These compounds have not been effective as anticancer agents, but some derivatives are promising.

Vinca rosea (commonly known as periwinkle) was originally known in folk medicine as a hypoglycemic agent. A further examination of this plant failed to substantiate this type of activity but did show anticancer activity. As a result, two alkaloids from this plant, Vinblastine and Vincristine, are presently used as anticancer agents.

The beneficial effect of an enzyme, L-asparaginase, as a cancer chemotherapeutic agent was seen in 1963. Malignant cells need L-asparagine for their normal growth. Normal mammalian cells do not. In malignant cells, L-Asparaginase converts L-asparagine to aspartic acid and ammonia. As a result, the normal amounts of L-asparagine are not available to the malignant cells, and their growth is suppressed. The L-Asparaginase can be adequately produced by *Escherichia coli.* Hence, a new type of anticancer agent became available.

Miscellaneous agents

As the knowledge of antibiotic drugs was increasing, synthetic drugs with chemotherapeutic properties were also being developed. In the nineteenth century, tuberculosis was a dreaded disease. The early treatment involved a long rest in a sunny tuberculosis sanatorium. Then, in 1944, Streptomycin was introduced in the treatment of the disease. Large doses were required, however, and toxicity appeared. Other compounds were required to augment its antitubercular activity and permit a smaller dose of streptomycin.

The effects of *p*-Aminosalicylic acid were observed on the metabolism of the tuberculosis organism in 1946. Soon afterward, it was used clinically in the treatment of the disease.

Thiacetazone was reported as an antitubercular agent in 1950; thilosemicarbazone derivatives of various other aldehydes were then prepared for antituberculosis testing. An intermediate in the preparation of isonicotinaldehyde, isonicotinic acid hydrazide, was tested for antitubercular activity. It was found to be a highly active compound with a wide margin of safety. This compound became known as Isoniazid and is still used as an antitubercular drug. Combination therapy of tuberculosis involving Streptomycin, *p*-Aminosalicylic acid, and Isoniazid has also been commonly used.

In 196l, the antitubercular activity of a series of ethylenediamine derivatives was observed, leading to the drug known as Ethambutol. In addition, a semisynthetic derivative of a macrocyclic antibiotic known as Rifampin was prepared in 1966. The bacterial inhibition of this substance against the tuberculosis organism was noted in 1968. Now, the combined use of Isoniazid, Ethambutol, and Rifampin constitutes a major weapon against tuberculosis.

The presence of a nitro group in any medicinal agent was formerly considered to cause toxicity, because many cases of methemoglobinemia were found among workers in munitions plants where trinitrotoluene was used. This compound was absorbed through the skin and reduced to aniline derivatives internally. In 1946, however, Doll introduced a series of nitrofuran drugs that showed good chemotherapeutic activity toward bacteria. In 1952, nitrofurantoin was introduced as an orally active urinary antiseptic, effective against both Gram-positive and Gram-negative organisms, and a new series of chemotherapeutic agents came into use.

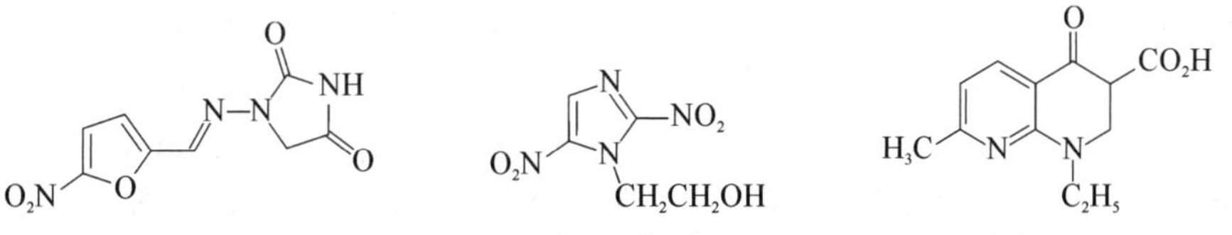

Nitrofurantoin　　Metronidazole　　Nalidixic acid

Metronidazole was patented in 1960 as a highly effective chemotherapeutic substance for the treatment of vaginal trichomoniasis. It was active orally and relatively free of side effects.

This drug has also opened new research concerning the nitroheterocyclic drugs.

The antiseptic activity of some quaternary ammonium salts was reported by Jacobs and Heidelberger in 1915. They introduced an alkyl group into the hexamethylenetetramine nucleus in an attempt to increase its antibacterial effectiveness. Quaternary amines that are more closely related to the present-day compounds were prepared by Hartmann and Kagi in 1928. Their series showed strong antiseptic activity.

The use of benzalkonium chloride as a detergent and germicidal agent was reported in 1935. Since that time, many cationic surfactants that are active as germicidal agents have been reported. These compounds can affect the enzymes and proteins of certain organisms, with the possible destruction of their cell walls.

Another drug that has shown good activity against urinary tract infections is Nalidixic acid. It was introduced in 1962 as a chemotherapeutic agent highly active against Gram-negative organisms. Even though the compound is inactive against Gram-positive organisms, it is important clinically and has opened up another area in the search for chemotherapeutic drugs.

Although past research has given us many good drugs for the treatment of infections, a number of parasitic infections remain for which no good chemotherapeutic agents are available. It is hoped that the search for agents against these infections, primarily tropical in their incidence, will be maintained to an extent commensurate with the occurrence of such infections in humans.

Words and Expressions

vesicant	*adj*. 起疱的(剂),腐烂性的	soybean	*n*. 大豆,黄豆
autopsies	*n*. [复]尸体解剖(检验)	germicidal	*adj*. 杀菌的
malignant	*adj*. 有恶意的,恶毒的	augment	*v*. 增大(加,长),扩大(张)
asparagine	*n*. 天冬酰胺	incorporate	*v*. (使)合并(并加)
respiratory	*adj*. 呼吸(作用)的	mustard-gas	*n*. 芥子气
mustard	*n*. 芥(禾),芥子气	miscellaneous	*adj*. 多方面的,其他的
detergent	*n*. 洗涤剂	tuberculosis	*n*. 结核(病),肺结核
mercaptopurine	*n*. 巯基嘌呤	antiseptic	*adj*.; *n*. 防腐(消毒,杀菌)的;防腐(抗菌,消毒)剂
munition	*n*.[复]军需(用)品,军火,弹药	spectrum/[复] spectra	*n*. 谱图
moldy/mouldy	*adj*. 发霉的,霉烂的		

Notes

① The azo linkage: linkage 通常指连接;联系;连锁;联动装置。字面上可以把该词组翻译为含氮的连接。通过上下文可知,本句是讲化学键,故翻译为含氮化学键。

② After the structures were determined, a reductive dehalogenation of Chlortetracycline was performed in 1953, producing Tetracycline. 单词 reductive 在制药、化学等专业领域里表示"还原的",不能翻译为普通英语"减少的"。dehalogenation 可拆分为以下几个词缀或词根:de、halogen、tion。de 作为前缀,具有"除去、离去"含义;halogen 表示卤素;dehalogen 就是除去卤素,专业术语为脱卤素;其动词为 dehalogenate,意为"脱卤素、去掉卤素";把词尾 e 去掉,加上名词后缀-tion 就构成名词 dehalogenation,含义是"脱卤素、脱卤作用"。Chlortetracycline 可拆分为 chlor(o)、tetra、cycl(o)、ine,中文分别表示氯代、四、环、药物,根据构词法,很容易获得该词中文含义"氯四环素",也叫作金霉素。

③ It was found to be a highly active compound with a wide margin of safety. 本句是比较常见的一种英语被动句表达方式,通常其谓语为"知道、了解、看见、认为、发现、考虑"等意义的动词,翻译时,通常可增加"大家、人们、我们、有人"等,译为主动句。margin of safety 简写为 MS,意为"安全系数、可靠性、安全边际、保险系数",本句中指安全系数。

Reading material B:

Apatinib: A Review in Advanced Gastric Cancer and Other Advanced Cancers

Unit 4
Microgels and Nanogels for the Delivery of Poorly Water-Soluble Drugs

While modern drug discovery approaches such as combinatorial chemistry, high-throughput screening, structural analysis of drug-target binding by X-ray diffraction, and molecular modeling have accelerated the identification of highly potent drug compounds, many of the structures identified have high melting points (T_m) or high octanol-water partition coefficients (log P, correlated with the lipophilicity of the drug) that limit the aqueous solubility of the drug and thus typically result in low bioavailability. Poorly water-soluble drugs, otherwise known as hydrophobic drugs, thus often require high doses to reach therapeutically relevant plasma concentrations and, subsequently, target tissue concentrations after administration, often leading to unwanted side effects and drug wastage.

The development of nanoparticle-based technologies for drug delivery has offered a new tool for improving the delivery of poorly soluble therapeutics. Many novel nanoparticle drug carrier systems have been studied and developed to improve the bioavailability of poorly water-soluble drugs by incorporating the drugs within the nanocarrier to facilitate targeted delivery. Due to their small size and large surface area, nanoparticles offer many benefits as vehicles for the delivery of poorly soluble drugs including (1) enhanced drug solubility; (2) improved permeation through size-selective barriers, including tight junctions of endothelial skin cells, deep into the lung following inhalation, and (in some cases) the blood-brain barrier; (3) targeted site-specific delivery of therapeutics to improve bioavailability; and (4) prolonged residence time within the body. Various nanoparticle systems (shown schematically in Figure 1-6) have been designed for the delivery of hydrophobic drugs, including dendrimers, micelles, solid lipid nanoparticles (SLNs), liposomes, and other polymeric nanoparticles comprised of either natural or synthetic materials. Table 1-1 summarizes the properties, advantages, and drawbacks of these different nanoparticlebased materials for hydrophobic drug delivery.

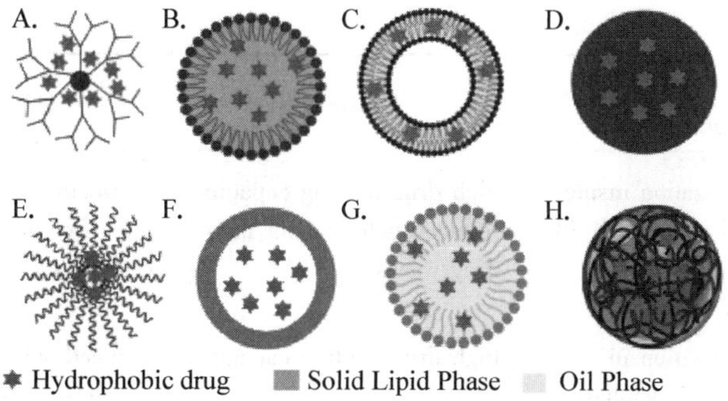

Figure 1-6 Nanoparticles used for the delivery of hydrophobic drugs: (A) dendrimers, (B) solid lipid nanoparticles, (C) liposomes, (D) polymeric nanoparticles, (E) micelles, (F) polymeric nanocapsules, (G) emulsions, and (H) microgels

Table 1-1 Properties, advantages, and drawbacks of various nanoparticles for hydrophobic drug delivery

Particle type	Hydrophobic drug loading strategy	Advantages	Drawbacks	Ref
dendrimers	void space between branches conjugation to surface functional groups	very small particle sizes possible very well-defined size and internal mass distributions	limited drug loading safety *in vivo* can be limited depending on composition	18
solid lipid nanoparticle (slns)	hydrophobic interactions/ solubilization in lipid phase	good safety profile easily scalable	can exhibit low drug loading capacity (particularly if solid is crystalline) often broader particle size distributions	16
liposomes	hydrophobic interactions with lipid bilayer	good safety profile/ biodegradability potential for dual hydrophobic/ hydrophilic drug delivery	potential for drug leakage or particle fusion due to bilayer dynamics limited volume for drug loading in bilayer often broader particle size distributions	9, 19
polymeric nanoparticles	hydrophobic interactions with water-insoluble polymer	simple preparation tunable degradation depending on polymer type and/or (if relevant) cross-linker chosen	stiff mechanics limit tissue penetration often broader particle size distributions	20
self-assembled block copolymer nanoparticles/ micelles	hydrophobic interactions with self-assembled hydrophobic core	good colloidal stability due to hydrophilic shell high drug loading capacity potential for degradation	stability of particle itself dependent on critical micelle concentration often broader particle size distributions (for polymeric micelles)	14, 21

continued

Particle type	Hydrophobic drug loading strategy	Advantages	Drawbacks	Ref
polymeric nanocapsules	solubilization inside liquid oil/solid lipid core inside polymeric shell	high drug loading capacity reduced potential for burst drug release	fabrication can be complex often broader particle size distributions	22
emulsions	solubilization in dispersed oil phase	high drug loading capacity	reduced stability in storage requires use of surfactants in dosage forms reduced *in vivo* stability	23
microgels	requires the inclusion of specific hydrophobic affinity domains	high interfacial hydration/ circulation times high colloidal stability deformable with potential to penetrate tight junctions even at higher particle sizes potential for narrow particle size distributions	low inherent affinity for hydrophobic drug loading	24

Emulsions, SLNs, self-assembled amphiphilic nanoparticles, liposomes, and polymeric nanoparticles, and (in most cases) dendrimers can inherently support hydrophobic drug delivery given that they each contain well-defined hydrophobic domains that enable partitioning-based physical loading of hydrophobic drugs. Microgels, defined as three-dimensional cross-linked networks of water-soluble polymers with sizes on the micrometer or smaller scale (with smaller microgels also referred to as nanogels), are in contrast inherently hydrophilic delivery vehicles; as such, they would typically enable significantly lower hydrophobic drug loading and less controlled hydrophobic drug release relative to more conventional drug delivery vehicles. However, microgels offer several other advantages that make their exploration for hydrophobic drug delivery attractive despite this inherent incompatibility: (1) The soft gel-like mechanics of microgels make them highly compressible, promoting the transport of drugs through tight junctions. (2) Multiple methods are available for fabricating microgels that result in very narrow and welldefined size distributions. (3) The high degree of interfacial hydration results in low protein deposition and thus typically extended circulation times, increasing the potential for tissue-specific targeting. (4) The flexible chemistry of microgels enables facile modification to enhance microgel affinity for specific drugs and/or the introduction of targeting ligands to enhance the potential for tissue-specific drug delivery. (5) Some synthesis methods can result in microgels with very narrow particle size distributions [polydispersity index (PDI) < 0.1], with potential benefits in terms of improving targeting of a given tissue. (6) Proper selection of precursor monomers can create stimuli-responsive microgels that can manipulate both drug uptake and drug release upon the addition of an external trigger. In addition, specific to hydrophobic drug delivery, the loading of poorly water-soluble drugs into the microgel causes them to collapse, in many cases enabling the formation of very small and highly stable nanoparticles that provide targeted drug delivery to tissues within the body.

Microgel characteristics such as size, charge, porosity, amphiphilicity, softness, and degradability can all be tuned by varying the type of water-soluble polymer(s)/monomer(s) used to fabricate the microgel, the type of cross-linker used (molecular weight, hydrophilicity, incorporation of degradable moieties), and the cross-linking density of the microgel, allowing for high customization of the microgel for specific applications. Notably, when microgels are fabricated based on stimulus-responsive polymers and/or cross-linkers, the resulting microgels are also stimulus-responsive to external cues from the environment such as changes in pH, temperature, redox conditions, enzyme concentrations, or the presence of external triggers, including light or magnetic fields. In such systems, these external cues result in conformational or structural changes in the nanogels facilitated by factors such as pH-sensitive ionization of acidic or basic functional groups, thermally responsive volume phase transitions that can alter microgel size (via swelling/deswelling transitions or cleavage of stimulus-responsive cross-links) and/or hydrophobicity (via thermally induced phase separation). Such switching capacity can be leveraged to facilitate localized microgel transport through tight junctions or promote site-specific drug release.

Despite these many advantages of microgels as drug delivery vehicles, the inherent mismatch between the hydrated bulk of a microgel and the hydrophobic drug being encapsulated poses a challenge toward achieving high encapsulation efficiency of hydrophobic drugs within hydrophilic microgel networks, at least using conventional affinity/partitioning-based loading strategies. This typically results in lower encapsulation efficiencies of hydrophobic therapeutics inside microgels relative to other carriers with a hydrophobic bulk phase, thus often requiring the delivery of a high amount of microgels to achieve a particular drug dose. Therefore, for the practical use of microgels for hydrophobic drug delivery, the identification of specific strategies to enable sufficient drug loading and subsequently control the rate of drug release is required. In this review, we highlight emerging approaches for modifying microgel compositions to enable efficient encapsulation of hydrophobic drugs. Furthermore, to address subsequent challenges in controlling the release of drug from hydrophobized microgels, we also discuss strategies to further modify microgels hydrophobized through these techniques to regulate hydrophobic drug release. We aim to highlight both the benefits and the drawbacks of microgels for hydrophobic drug delivery to inform both microgel researchers aiming to synthesize hydrophobized microgels with optimal properties for drug delivery as well as drug formulators aiming to leverage the beneficial properties of microgels for hydrophobic drug delivery.

引自:DAVE R, RANDHAWA G,KIM D, et al. Microgels and Nanogels for the Delivery of Poorly Water-Soluble Drugs[J]. Molecular Pharmaceutics, 2022, 19(6): 1704-1721.

Words and Expressions

microgel	*n.* 微凝胶	nanogel	*n.* 纳米凝胶
lipophilicity	*n.* 亲脂性	hydrophobic	*adj.* 疏水的
nanoparticle	*n.* 纳米粒子	inhalation	*n.* 吸入
dendrimers	*n.* 树枝状聚合物	micelle	*n.* 微胶粒
liposome	*n.* 脂质体	emulsion	*n.* 乳剂
flexible	*adj.* 灵活的	charge	*n.* 电荷
amphiphilicity	*n.* 双亲性	degradability	*n.* 降解性
monomer	*n.* 单体	conformational	*adj.* 构象的
swelling	*n.* 肿胀	encapsulation	*n.* 封装
combinatorial chemistry	组合化学	tissue-specific	组织特异性
high-throughput screening	高通量筛选	polydispersity index	多分散指数
melting points	熔点	drug uptake	药物吸收
the blood-brain barrier	血脑屏障	redox condition	氧化还原条件

Notes

① Drug delivery vehicles
药物递送载体

② High octanol-water partition coefficients (log *P*, correlated with the lipophilicity of the drug)
油水分配系数或疏水性参数或疏水系数(与药物的亲脂性有关)

③ Solid lipid nanoparticles
固体脂质纳米粒

④ Poorly water-soluble drugs, otherwise known as hydrophobic drugs, thus often require high doses to reach therapeutically relevant plasma concentrations and, subsequently, target tissue concentrations after administration, often leading to unwanted side effects and drug wastage.
水溶性差的药物,又称疏水性药物,往往需要高剂量才能达到治疗的血药浓度和目标组织浓度。疏水性药物常常导致副作用和药物浪费。

Exercises

1. Put the following into English:

组合化学　　亲脂性　　血脑屏障　　药物吸收　　降解性

2. Put the following into Chinese:

a. Hydrophobization of the microgel is likely to at least partially compromise the inherent advantages of microgels in suppressing protein adsorption relative to other types of nanoparticles.

b. In our view, provided the interface of the microgel retains the beneficial water binding/low interfacial energy property of a microgel, the ability to access much smaller particle sizes while retaining the benefits of colloidal stability and compressibility has significant potential to address challenges with traversing biological barriers such as the blood-brain or blood-retina barrier, among others.

Unit 5
Computer Aided Drug Design

Introduction

Drug discovery is a capital and time-intensive process which is aimed at developing new drug candidates. One way of achieving this has been through the aid of computational means in the pre-clinical phase of drug discovery. Computer-aided drug design (CADD) can be defined as computational approaches that are used to discover, develop and analyse drug and active molecules with similar biochemical properties. Homology modelling, molecular docking, virtual screening (VS) or virtual high-throughput screening (vHTS), quantitative structure-activity relationship (QSAR) and three-dimensional (3D) pharmacophore mapping generally, are the main constituents of CADD. Among these techniques, it seems that virtual screening is the major contributor to CADD and it has become somewhat a proven and well-appreciated computational method, that stands as a contemporary to the experimental high-throughput screening for hit identification and optimization.

To date, the discovery process of more than 70 commercialised drugs included some form of computational technique (Table 1-2), significant enough for being mentioned in literature. Note that in most cases the initial drug lead (not the final commercial drug) was discovered with the aid of some CADD technique. In our own experience, however, scepticism about the contribution of CADD still exists. SciFinder® was systematically and exhaustively searched, using three sets of keywords consistently for this review, relating to virtual screening, software programs, computational studies and a specific year. From the analysis of literature data, it appears to us that the steady increase of commercial drugs via the CADD route may be linked to a growth of molecular modelling/computational chemistry research, as well the current trend of increased multidisciplinary research. This hypothesis will subsequently be tested. The current review was based on a hybrid approach building on two previous review papers (2006 and 2016) whose concepts were combined, modified and refined. For the six-year period under study, a total of 618 publications were considered for evaluation on which this review is based. The data for all the studies considered in this review, and the breakdown of the various components of interest per individual year between 2015 and 2020 is presented and summarised in the supplementary section.

Table 1-2 Commercial drugs that made use of CADD during the discovery process

Inhibitor Name	Protein Target	Computational contribution to drug discovery	Approval
Captopril (Capoten)	Angiotensin converting enzyme (ACE)	LBDD and Structure-activity relationship (SAR) and SBDD and Ligand-based drug design	1981
Norfloxacin (Noroxin)	Topoisomerase Ⅱ, Ⅳ	SBVS, LBDD and QSAR Modelling	1986
Flurbiprofen	Cyclooxygenase-2	Molecular Docking	1988
Imatinib	Tyrosine kinase	SBDD (chemical libraries were screened for inhibitors against Bcr-Abl tyrosine kinase)	1990
Epalrestat	Aldose Reductase	MD and SBVS	1992
Cladribine	Adenosine deaminase	SBDD (VS and docking)	1993
Dorzolamide (Trusopt)	Carbonic anhydrase (CA) Ⅱ	Fragment-based Screening, SBDD and ab initio calculations	1994
Losartan (Cozaar)	Angiotensin Ⅱ receptor	LBDD (traditional screening and ligand-based optimization)	1995
Saquinavir (Invirase)	HIV-1 protease	SBDD (Transition-state mimetic concept)	1995
Indinavir (Crixivan)	HIV-1 protease	SBDD (Transition-state mimetic concept guided by molecular modelling and X-ray crystal structure)	1996
Ritonavir (Norvir)	HIV-1 protease	SBDD, LBDD, SAR and Lead Optimization (Transition-state mimetic concept)	1996
Delavirdine	HIV reverse transcriptase	SBDD (Virtual screening against HIV-1 RT) and LBDD (Lead optimization and SAR)	1997
Nelfinavir (Viracept)	HIV-1 protease	LBDD (Iterative protein cocrystal structure analysis and lead optimization), iterative SBDD and Transition-state mimetic concept	1997
Efavirenz	Non-nucleoside reverse transcriptase	SBDD (Screening compounds against HIV-1 RT via computational dissimilarity analysis and lead structure optimization)	1998
Tirofiban (Aggrastat)	Integrin (GP) Ⅱb/Ⅲa and Fibrinogen receptor	Ligand-based design using RGD-binding motif and Ligand-based pharmacophore screening	1999
Amprenavir (Agenerase)	Antiretroviral (HIV-1) protease	Combined SBDD, LBDD, Protein Modelling and Molecular Dynamics	1999
Oseltamivir	Influenza A and B neuraminidase	Rational drug design utilizing available high-resolution X-ray crystal structures and the transition state analogue	1999
Zanamivir (Relenza)	Influenza Neuraminidase	SBDD (Computer-assisted modelling of the active site)	1999
Lopinavir	HIV-1 protease	SBDD (Transition-state mimetic concept) [3D modelling and docking. Energy minimization using DISCOVER CVFF force field]	2000

continued

Inhibitor Name	Protein Target	Computational contribution to drug discovery	Approval
Eprosartan	Angiotensin Ⅱ receptor	SBDD (computerized molecular modelling overlap of Angiotensin Ⅱ with the structure of S-8308) and LBDD (Pharmacophore modelling)	2001
Eptifibatide	Glycoprotein Ⅱb/Ⅲa	Peptide-based (barbourin) design	2001
Imatinib mesylate	Abl tyrosine kinase	SBDD	2001
Valsartan (Diovan)	Angiotensin Ⅱ receptor	Superimposition of energy-minimized conformation and QSAR	2002
Atazanavir	HIV-1 protease	Computational mapping of protein binding sites and docking of the ligands	2003
Enfuvirtide (Fuzeon)	HIV-1 protease	Homology Modelling	2003
Fosamprenavir	HIV-1 protease	Structure-based design	2003
Gefitinib	GFRv tyrosine kinase	SBVS	2003
Zolmitriptan (Zomig)	5-hydroxytryptamine (5-T)$_{1B/1D/(1F)}$ receptor	LBDD and Pharmacophore Modelling	2003
Erlotinib	EGFR kinase	SBVS	2005
Tipranavir	Nonpeptidic HIV-1 Protease	(Combination of a Monte Carlo based automatic docking program and X-ray crystallography, lead optimization)	2005
Darunavir	Nonpeptidic HIV-1 protease	Combined SBDD and LBDD	2006
Sunitinib (Sutent)	VEGF-R2 kinase	Structure-activity relationships (SARs) and Homology Modelling	2006
Aliskiren (Tekturna)	Angiotensinogen	SBDD and Docking	2007
Ambrisentan	Endothelin-A receptor	SBDD (Docking), FBDD and Virtual Screening	2007
Maraviroc	CCR5/gp120	SB Molecular Modelling	2007
Nilotinib (Tasigna)	Bcr-Abl mutant	SBDD	2007
Raltegravir	HIV-1 integrase	Combining MD with flexible-ligand docking	2007
Tomudex	Thymidylate synthase	SBDD	2009[#]
Boceprevir	Hepatitis C virus (HCV)	SBDD approach (Transition-state mimetics and SAR) guided by X-ray crystal structures and SAR optimization	2011
Crizotinib (Xalcori)	ALK and ROS1	SBDD and SAR	2011
Rilpivirine (Edurant)	Non-nucleoside reverse transcriptase (NNRT)	SBDD (homology modelling, compound screening, docking, mechanistic studies and transition state isostere modelling)	2011

continued

Inhibitor Name	Protein Target	Computational contribution to drug discovery	Approval
Rivaroxaban	Factor Xa	HTS, SBDD and Virtual SAR	2011
Telaprevir	NS3/4 A protease	Substrate-Based Inhibitor Design and Structure-Based Inhibitor Optimization	2011
Dolutegravir	HIV-1 Integrase	PBDD (two-metal binding pharmacophore structural based design)	2013
Saroglitazar (Cevoglitazar)	PPAR	Combined virtual screening of 3D databases, SBDD and Pharmacophore Modelling	2013 *
Grazoprevir (Zepatier)	NS3/4 A protease	Molecular Modelling and Docking-derived approach	2016
Lifitegrast	LFA-1/ICAM-1	Structure-based rational design	2016
Rucaparib (Zepatier)	Poly (ADP-ribose) polymerase (PARP-1)	Ligand-based molecular modelling	2016
Venetoclax	Bcl-2/(BAX/BAK)	Rational design for BCL-2	2016
Acalabrutinib	Bruton's tyrosine kinase	SAR, SBDD and Docking	2017
Betrixaban	Serine protease Factor Xa (fXa)	Molecular Docking	2017
Brigatinib (Alunbrig)	ALK	Docking and Homology Modelling	2017
Copanlisib Hydrochloride	Phosphoinositide 3-kinase (PI3K)	SBDD (X-ray crystallography and Docking) and LBDD (based on lead scaffold)	2017
Vaborbactam (Vabomere)	β-Lactamase	Docking and MD	2017
Abemaciclib	Cyclin-dependent kinase	Structure-activity relationship studies in conjunction with structure-based design	2018
Apalutamide	Androgen receptor inhibitor	SBDD and SAR	2018
Dacomitinib	Oral kinase	Combined FBDD and SBDD	2018
Duvelisib	PI3K Kinase	SBDD (Molecular docking, virtual screening) and LBDD (lead optimization and SAR)	2018
Glasdegib Maleate	Hedgehog pathway	SAR	2018
Ivosidenib	Isocitrate dehydrogenase-1 (IDH1)	LBDD coupled with broad SAR profiling and modification	2018
Larotrectinib Sulphate	Tropomyosin-related kinase	LBDD with SAR and crystal-binding mode similarity	2018
Lorlatinib	Tyrosine kinase	SBDD and physical property-based optimization	2018
Talazoparib Tosylate	Poly (ADP-ribose) polymerase PARP	SBDD, SAR and Lead Optimization	2018
Darolutamide	Androgen receptor	SBDD (Docking and MD)	2019

continued

Inhibitor Name	Protein Target	Computational contribution to drug discovery	Approval
Entrectinib	Tyrosine kinase inhibitor	SBDD and SAR	2019
Erdafitinib	FGFR tyrosine	Combined FBDD and SBDD	2019
Fedratinib Hydrochloride	Tyrosine kinase	SBDD (Virtual screening and Molecular Docking)	2019
Selinexor	Nuclear export	SBDD (consensus induced fit docking)	2019
Zanubrutinib	Bruton's tyrosine kinase inhibitor	Combined FBDD and SBDD	2019

The data presented in the table was compiled from various literature sources [5, 16, 182, 183, 290, 292, 295, 303, 304, 306, 346-353]. * approved by Drug Controller General of India (DCGI). # approved by European Medicines Agency (EMA). Data regarding the computational contribution was not always readily presented, thus some drugs were left out. Some protein targets are presented in abbreviated form. HTS—High Throughput Screening; MD—Molecular Dynamics; LBDD—Ligand-Based Drug Design; PBDD—Pharmacophore-Based Drug Design; SBDD—Structure-Based Drug Design; SBVS—Structure-Based Virtual Screening; SAR—Structural Activity Relationship; QSAR—Quantitative Structural Activity Relationship.

A myriad of virtual compound libraries and databases can be utilized for 1D, 2D and 3D compound screening. Several methodological protocols are also available offering a range of similarities and contrasts, resulting in favoured application of some methods over others. Other computational and non-computational techniques in the pre-clinical drug discovery pipeline complement virtual screening and they are used to determine the affinity, selectivity, stability and bioavailability of the lead compounds. Virtual screening is a relatively efficient and inexpensive tool applied in drug discovery for the systematic search of novel small molecules with biological activity usually against a target protein molecule. The backbone of this computational method rests upon the outstanding improvements in computing algorithms, a dramatic increase in the processing power of computers, the vast knowledge of structural and physico-chemical properties of compounds in libraries and databases, and the increased knowledge of the structural and functional properties of protein molecules. The three-dimensional visualizations involved in virtual screening allow for more refined insight and ease of manipulation. This method can be applied to screen for chemical compounds (natural and synthesized), peptides or proteins. There are two main facets to virtual screenings which are ligand-based (LBVS) and structure-based (SBVS). From these two, a host of other virtual screening methods crop up, such as: accelerated free energy perturbation-based, chemical genomics-based, docking-based, fragment-based, homology model-based, hybrid, inverse, pharmacophore-based, QSAR-based, conformal prediction based, retrospective, shape-based, and similarity-based virtual screening etc. Virtual screening is also important for drug repurposing or repositioning, thus enabling the quick characterization and optimization of novel drug candidates, which consequently speeds up drug design and development.

Molecular docking has been widely used within virtual screening to assist in streamlining the search especially where a protein 3D structure is available. There are essentially three categories of docking, i. e., ensemble, induced work and lock and key docking. Docking methodologies include rigid ligand and rigid receptor docking; and flexible ligand and rigid receptor docking and flexible ligand and flexible receptor docking. Various docking and scoring software programs are available as in-house or opensource applications for use to that effect, employing different algorithms and functions. Docking algorithms include genetic, incremental construction, LUDI, matching, multiple-copy simultaneous search (MCSS), and Monte Carlo algorithms. The aim of docking is to predict the bound conformation of a ligand within the binding site of a receptor. Scoring functions such as descriptor-based, empirical, force-field based, and knowledge-based are used to determine the best binding conformations.

However, docking techniques have several well-documented shortcomings. It is known that the majority of docking software programs employ force field calculations that make estimations of the binding energy, guided by quantum mechanics (QM) and experimental data. However, accurate binding energies can only be determined from methods involving ab initio calculations, including DFT as well as molecular dynamics simulations. In order to simplify the calculations, most docking software programs tend to remove hydrogens (protons) of the molecules under study (enzyme and inhibitors), thus vital information is excluded leading to inaccuracies. Further information can be obtained on these preliminary results through MD simulations, in which aspects such as protonation and solvation are considered. A few mentionable examples of docking shortcomings are summarised as follows: inadequate resolutions of crystallographic targets; flexibility issues (both structural and conformational); the simplifications and assumptions in scoring functions that negatively impact on accuracy; limitations in accurately accounting for hydrogen bonding, directional interactions, solvent and entropic effects. These issues have been evaluated and discussed in several reviews and studies.

Machine learning and deep learning have also been applied to virtual screening or as a refinement to the virtual screening process. These have been used as standalone tools, in combination with other virtual screening methods or as a comparison/benchmarking for virtual screening applications. Machine learning and deep learning have been applied in various virtual screening studies to enhance the effectiveness of similarity searching and data mining in ligand-based virtual screening; to improve the scoring functions for structure-based virtual screening; and performance evaluation. One area of CADD that has benefited from advances in these two approaches is de novo drug design (DNDD). This is an iterative computational approach in which novel lead molecules are designed from scratch with the basic guidance of constraints via computational algorithms. Detailed aspects regarding de novo drug design have been discussed comprehensively before. Aspects relating to machine learning, deep learning and DNDD will not be evaluated in this current review.

Molecular dynamics simulations are important as a guide in improving the binding properties and efficacy of the lead compounds through the use of Newtonian mechanics. These calculations are superior to docking and several studies have also applied molecular dynamics simulations to complement virtual screening. MD simulations are essential in providing the dynamics of the protein structure as well as producing refined conformations due to the possibility of generating multiple conformer snapshots. Furthermore, ab initio methods and density functional theory can be used to enhance virtual screening by revealing the intrinsic atomistic-electronic interactions between protein and ligand molecules. In simple terms, MD simulations produce more reliable binding affinity results, therefore it is advisable that researchers apply the technique to confirm the validity of their docked results.

Ultimately, experimental evaluation/calibration is typically used for hypothesis testing and to serve as a benchmark for validating the computational model. There is however no standardized approach to virtual screening, as a result the method is applied as befits the purpose. Various studies have been performed with virtual screening, and numerous combinations of the aforementioned computational techniques have been used to complement virtual screening in drug discovery. The drug discovery effort should fundamentally lead to either commercial or public health success where tangible results can be identified in this case, in the form of approved drugs. Since the inception of computational techniques several drugs which are currently on the market have been discovered with the aid of CADD and several others are still in the pipeline. This has made virtual screening a crucial technique in drug discovery.

In this review we will evaluate the current trends and biases in pre-clinical drug discovery in the form of computer-aided drug design (CADD) specifically with respect to virtual screening. We will analyse the current use of virtual compound libraries, databases and present an overview of the recent trends and applications in virtual screening. We hope to provide insight on the various techniques applied together with virtual screening in the last six years. A general overview of the available drugs whose discovery pipeline involved various CADD techniques including homology modelling, molecular docking, pharmacophore mapping, QSAR and virtual screening will be provided.

引自:SABE V T, NTOMBELA T, JHAMBA L A, et al. Current trends in computer aided drug design and a highlight of drugs discovered via computational techniques: A review. European Journal of Medicinal Chemistry, 2021, 224, 113705.

Words and Expressions

homology	*n.* 同源	pharmacophore	*n.* 药效团
initial	*adj.* 最初的	scepticism	*n.* 怀疑论
breakdown	*n.* 数字明细	pipeline	*n.* 管线
affinity	*n.* 亲和力	bioavailability	*n.* 生物利用度
algorithms	*n.* 算法	virtual	*adj.* 虚拟的
manipulation	*n.* 操作	peptide	*n.* 肽,缩氨酸

protein	*n.* 蛋白质	enzyme	*n.* 酶
rigid	*adj.* 刚性	ligand	*n.* 配体
flexible	*adj.* 柔性	entropic	*adj.* 熵
in-house	*adj.* 内部	hydrogen	*n.* 氢
conformation	*n.* 构象	dynamic	*n.* 动力学

Notes

① Quantitative structure-activity relationship (QSAR)
定量构效关系

② It is known that the majority of docking software programs employ force field calculations that make estimations of the binding energy, guided by quantum mechanics (QM) and experimental data.
众所周知,大多数对接软件以量子力学和实验数据为指导,采用力场计算来对结合能进行估算。

③ In order to simplify the calculations, most docking software programs tend to remove hydrogens (protons) of the molecules under study (enzyme and inhibitors), thus vital information is excluded leading to inaccuracies.
为简化计算,大多数对接软件都倾向于去除所研究分子(酶和抑制剂)的氢原子(质子),因此排除了导致不准确的重要信息。

④ One area of CADD that has benefited from advances in these two approaches is de novo drug design (DNDD).
从这两种方法中获益的计算机辅助药物分子设计领域之一是全新药物设计。

Exercises

1. Put the following into English:
生物利用度　　柔性　　构象　　亲和力　　药效团

2. Put the following into Chinese:

a. From the two fundamental virtual screening approaches (ligand-based and structure-based) emanates a host of other virtual screening techniques namely accelerated free energy perturbation-based, chemical genomics-based, docking-based, fragment-based, homology model-based, hybrid, inverse, pharmacophore-based, QSAR-based, conformal prediction based, retrospective, shape-based and similarity-based virtual screening etc.

b. The impact of CADD in pre-clinical drug development is profound and its application is increasing with the advances in virtual screening and molecular docking.

Reading material A:

Report on the Sunway TaihuLight System

The 93 petaflop Sunway TaihuLight is installed at the National Supercomputing Centre in Wuxi. At its peak, the computer can perform around 93, 000 trillion calculations per second.It is twice as fast and three times as efficient as the previous leader Tianhe-2, also from China, said Top500 which released the new list on Monday. Its main applications include advanced manufacturing, weather forecasting and big data analytics, wrote Jack Dongarra in a paper about the new machine. It has more than 10.5 million locally-made processing cores and 40, 916 nodes and runs on a Linux-based operating system. For the first time since the list began, China has overtaken the U.S. with 167 computers in the top 500 while the U.S. has 165. "Considering that just 10 years ago, China claimed a mere 28 systems on the list, with none ranked in the top 30, the nation has come further and faster than any other country in the history of supercomputing," said Top500. The U.S. has four supercomputers in the top 10 of the Top500 list, while China has two which currently occupy the top two places. The other positions in the top 10, published twice a year, are occupied by machines from Japan, Switzerland, Germany and Saudi Arabia. "As a computer scientist it's difficult writing software that can take advantage of and control large numbers of computer cores," said Professor Les Carr from the University of Southampton. "This is why supercomputers are restricted to specialised applications—you need very specialised computing needs to take advantage of them. "They are like extremely high-spec Grand Prix racing cars—they are fantastic for racing on circuits but they're not great for travelling from London to Edinburgh."

Reading material B:

Natural Product Drug Discovery in the Artificial Intelligence Era

Reading material C:

AI CIANTS can Benefit Other Nations

Unit 6
Cancer Drug Resistance Related microRNAs: Recent Advances in Detection Methods

Cancer, as a general term for malignant tumors, seriously endangers human health and life, thus becoming one of the leading causes of the global death. Currently, common treatments for cancer mainly consist of chemotherapy, radiotherapy and surgery, of which chemotherapy is the first-line treatment for some cancers, including small cell lung cancer, lymphoma and leukemia in order to control malignant tumor growth and prolong patient survival. And, for nasopharyngeal carcinoma, liver cancer, breast cancer and other entity tumors, chemotherapy can be employed for local treatment before surgery or radiotherapy, or it can serve as an adjuvant therapy to clear away residual nodules after surgery to avoid recurrence. However, cancer drug resistance becomes a challenge in cancer therapy, which will hinder effective treatment and result in a poor prognosis. Drug resistance is the phenomenon in which a disease has resistance to drug treatment, which is classified as innate resistance and adaptive resistance. Innate resistance is the inherent property of tumors with natural unresponsiveness to drug treatment, while adaptive resistance is acquired after the disease has been treated with drugs for a period of time. Cancer drug resistance caused by various factors will promote tumor cell proliferation and migration, and stimulate the occurrence and development of cancers. It is noted that patients who have been cured may also have cancer recurrence due to drug resistance. Therefore, it is imperative to identify the mechanism of drug resistance and devise effective treatment strategies to achieve great treatment outcomes.

MicroRNAs (miRNAs) are short non-coding RNAs with lengths of about 19–25 nucleotides, which can negatively regulate the expression of specific target genes through post-transcriptional interference of mRNA via binding to the 3'UTR of mRNA. miRNAs play an important role in various regulatory mechanisms of organisms. miRNAs are utilized as regulatory factors over a broad spectrum of cancer pathogenesis, which involves a variety of physiological and pathological processes. And they can also be used as oncogenes or tumor suppressors to regulate tumor progression or affect tumor invasion. Importantly, miRNAs are related to cancer drug resistance through targeting genes and proteins associated with cell proliferation, cycle and apoptosis, regulating drug metabolic enzymes and transporters, modulating DNA damage repair, and impacting on the epithelial to mesenchymal transition (EMT). In parallel, regulation has a twofold effect: the dysregulation of miRNAs will promote drug resistance, and conversely, effective regulation of miRNAs will reduce drug resistance. In addition, miRNAs secreted from exosomes, as signal molecules between tumor cells and the microenvironment, can be transported to various parts of the body, resulting in promoting and conferring drug resistance or sensitivity to treatments.

As miRNAs are of great value for regulating cancer drug resistance and other physiological processes, there is great demand for the development of effective methods for miRNA detection. However, miRNAs possess the characteristics of short sequences and high homology, which makes the detection of miRNAs challenging. Dating back to 1993, northern blotting was the first method to detect miRNAs, subsequently, microarrays and the reverse transcription-quantitative polymerase chain reaction (RT-qPCR) were utilized as miRNA detection tools. Although these traditional approaches have further improved our understanding of the role and function of miRNAs, they suffer from some limitations. The northern blotting method is complex, has poor efficiency and requires radiolabeling, which can introduce contamination. The microarray method lacks detection sensitivity and repeatability. And, RT-qPCR as a gold standard method of miRNA detection has great detection limits, but it calls for a complex temperature regulation program and has a high cost of use. More than a decade later, to overcome these limitations, some advanced methods with great detection efficiency, high sensitivity and low cost were developed gradually. In 2006, Jonstrup et al. first applied rolling cycle amplification (RCA), an isothermal amplification method, to miRNA detection. Thereafter, graphene oxide (GO), gold nanoparticles and other nanomaterials were also employed for detection. Recently, chromatography, mass-spectrometry and other methods were devised.

Herein, biogenesis of miRNAs will be introduced. In the nucleus, primary miRNAs (pri-miRNAs) are transcribed by specific genes under the action of RNA polymerase Ⅱ and then cut by the Drosha-DGCR8 complex into the pre-miRNAs. Next, precursor miRNAs (pre-miRNAs) are transported into the cytoplasm by Exportin-5 and converted into miRNA duplexes by the Dicer-TRBP-PACT complex. Finally, single-stranded miRNAs are formed under the action of helicase. The relationships between miRNAs and cancer drug resistance are also summarized. There is a correlation between miRNAs and cancer drug resistance for regulating cell proliferation, cycle and apoptosis, adjusting drug metabolism and transport, conducting DNA damage repair and controlling the EMT. Furthermore, miRNA detection methods, including isothermal amplification methods, nanomaterial-related methods, and other technique-based methods will be discussed. This review will facilitate the development of the non-invasive diagnosis of cancer drug resistance, as well as the inhibition of drug resistance by miRNAs (Scheme 1-1).

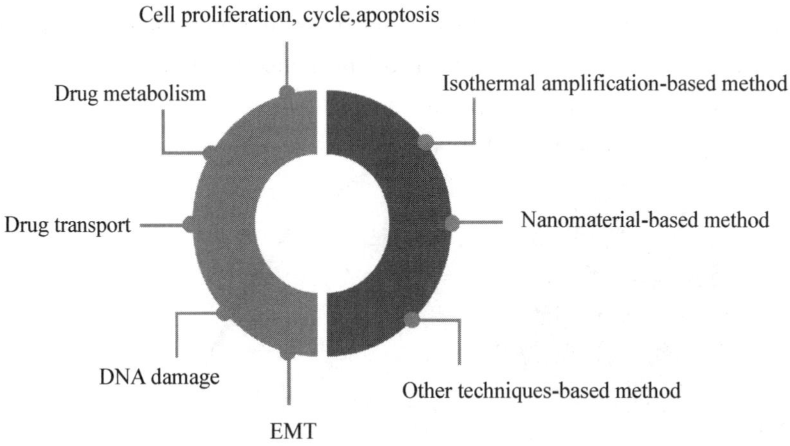

Scheme 1-1 The mechanisms of cancer drug resistance and the methods for miRNA detection

Biogenesis of microRNAs

Generally, mature miRNAs with certain functions are generated from primary transcripts that undergo a series of biogenesis processes. miRNAs with genes located in intron regions have specific promoters or share promoters with host genes, while genomic miRNAs, as a portion of the common transcript, are transcribed directly. The majority of miRNAs are produced through a canonical biogenesis pathway (Figure 1-7). First of all, miRNA genes are transcribed to large polyadenylated RNAs named pri-miRNAs, which have the structure of a stem loop with a length of hundreds to thousands of nucleotides (nts) by RNA polymerase Ⅱ. Subsequently, primiRNAs are identified and cut by the Drosha-DGCR8 complex (so-called microprocessor) in the nucleus and converted into pre-miRNAs, which are roughly 70-90 nts with a small hairpin structure. During this process, the two subunits of the microprocessor have different functions. Drosha (a protein complex containing the RNase Ⅲ dsRNA-specific endoribonuclease nuclear type Ⅲ) serves as a catalyst, while DGCR8 (DiGeorge syndrome critical region 8), a dsRNA binding protein, is not only required to stabilize Drosha via interactions, but also acts as a molecular ruler to determine the cleavage site of Drosha. Exportin-5, a Ran-GTP-dependent dsRNA-binding intranuclear protein, transports pre-miRNAs from the cell nucleus to the cytoplasm. The transport specificity is determined by the structure of miRNAs, such as the lengths of the 3′ protruding fragment and stem. Furthermore, exportin-5 plays a vital role in protecting pre-miRNAs from nuclear degradation. After transport, pre-miRNAs are converted into miRNA duplexes through cleavage mediated by the Dicer-TRBP-PACT complex in the cytoplasm; Dicer is an Rnase Ⅲ-type cytoplasmic endoribonuclease, TRBP is a Tar RNA-binding protein and PACT is a protein activator. Finally, the miRNA duplex is unwound by helicase into two individual single-stranded miRNAs. Typically, either strand of ds-miRNA is recognized as mature ss-miRNA and acts as the core region that is loaded into the Argonaute (AGO) protein complex to form the RNA-induced silencing complex (RISC). However, existing research shows that the single-stranded miRNA with a less stable base pair at the 5' end as the mature miRNA is assembled into the RISC to induce gene silencing; nevertheless, the one with a more stable base pair is degraded.

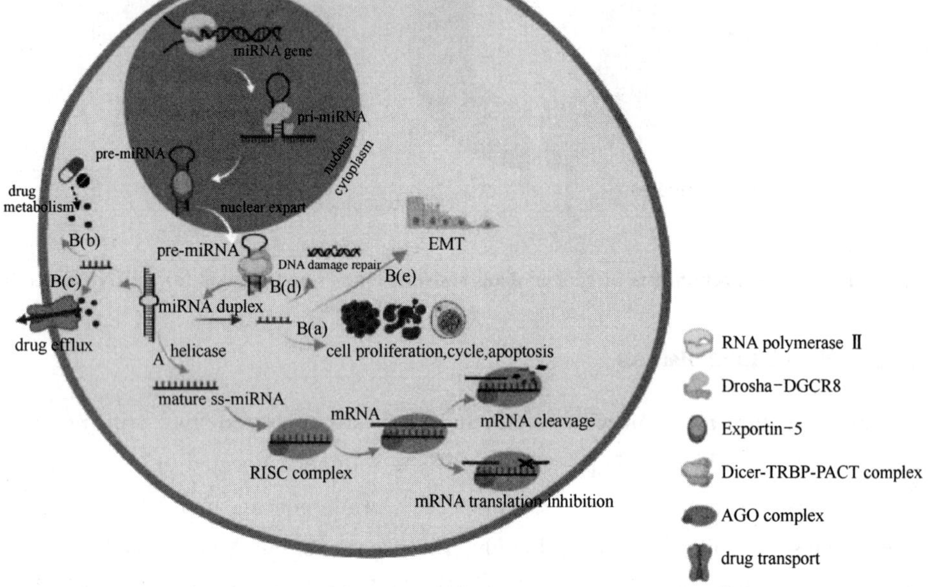

Figure 1-7 The biogenesis of miRNAs

Loading of mature miRNA into the RISC complex indicates completion of miRNA biogenesis. Therefore, the AGO protein family, a core element of the RISC with four subfamilies related to miRNA pathways (AGO1-4), has RNA binding domains for recognizing and binding target RNAs, and AGO2 in particular contains an active RNase binding domain with a splicing function. Once assembled into the RISC, mature miRNAs, assisted by the AGO protein, bind to the target genes by discerning a sequence of 2-8 nts in length on the 3' UTR of target mRNAs, also called "the seed region". According to the completeness of the mRNA base pair complementarity, miRNAs will mediate two different mechanisms. If miRNAs are entirely complementary with mRNAs, mRNAs will be cleaved. And miRNAs will lead to translational inhibition when miRNAs and mRNAs are insufficiently complementarity.

However, research has suggested that this canonical pathway has exceptions; some special miRNAs achieve biogenesis in a non-canonical manner. Peculiar miRNAs are generated from mirtrons through the spliceosome machinery. Mirtrons, inherent segments of mRNAs, first found in Caenorhabditis elegans and Drosophila melanogaster, are processed into pre-miRNAs independent of the Drosha-DGCR8 complex. Besides, small nuclear RNAs can also lead to miRNAs, which bypass processing by the Drosha-DGCR8 complex.

引自:Hu X Y, SONG Z, YANG Z W, et al. Cancer drug resistance related microRNAs: recent advances in detection methods[J]. Analyst, 2022, 147(12): 2615-2632.

Words and Expressions

radiotherapy	*n.* 放射治疗	chemotherapy	*n.* 化疗
leukemia	*n.* 白血病	lymphoma	*n.* 淋巴瘤
proliferation	*n.* 增殖	microRNAs	*n.* 微小核糖核酸
cycle	*n.* 循环	migration	*n.* 迁移
exosome	*n.* 外泌体	mechanism	*n.* 机制
blotting	*n.* 印迹	pathological	*adj.* 病理学的
polymerase	*n.* 聚合酶	nucleotide	*n.* 核苷酸
helicase	*n.* 解螺旋酶	apoptosis	*n.* 细胞凋亡
radiolabeling	*n.* 放射性标记	microenvironment	*n.* 微环境
cytoplasm	*n.* 细胞质	homology	*n.* 同源
entity tumors	实体瘤	breast cancer	乳腺癌
drug resistance	耐药性	innate resistance	天然耐药
tumor invasion	肿瘤侵袭	malignant tumors	恶性肿瘤
short sequence	短序列	adaptive resistance	适应性耐药
nasopharyngeal carcinoma	鼻咽癌		

Notes

① Importantly, miRNAs are related to cancer drug resistance through targeting genes and proteins associated with cell proliferation, cycle and apoptosis, regulating drug metabolic enzymes and transporters, modulating DNA damage repair, and impacting on the epithelial to mesenchymal transition (EMT).

重要的是,miRNAs 通过细胞增殖、周期凋亡,调控药物代谢酶和转运体,调节 DNA 损伤修复、影响上皮细胞间充质转化等途径与癌症耐药相关。

② Dating back to 1993, northern blotting was the first method to detect miRNAs, subsequently, microarrays and the reverse transcription-quantitative polymerase chain reaction (RT-qPCR) were utilized as miRNA detection tools.

从 1993 年开始,Northern 印迹是检测 miRNA 的第一种方法,随后基因芯片和逆转录-定量聚合酶链反应被用作 miRNA 的检测工具。

③ DGCR8 (DiGeorge syndrome critical region 8)

迪乔治综合征危象区基因 8

④ Rolling cycle amplification(RCA)

滚环扩增

⑤ RNA-induced silencing complex(RISC)

RNA 诱导沉默复合物

Exercises

1. Put the following into English:

   耐药性　　乳腺癌　　增殖　　核苷酸　　微环境

2. Put the following into Chinese:

a. It is worth mentioning that drug resistance regulatory miRNAs can synergistically suppress tumor growth and progression with chemotherapy drugs, which enhances the therapeutic effect of drugs and reverses drug resistance in a way.

b. We focus on the relationship between miRNAs and cancer drug resistance, and discuss the roles of miRNAs in different drug resistance mechanisms.

Unit 7

Progress Towards a Clinically-Successful ATR Inhibitor for Cancer Therapy

The integrity of human DNA is constantly subject to damage either by exogenous exposure to radiation or genotoxic agents, or by endogenous reactive and oxidative by-products of normal cellular metabolisms. This damage, if not repaired or incorrectly repaired, becomes lethal to the cell or organism. Constant and efficient repair of this DNA damage is therefore essential and biologically significant in preventing cellular death and many human diseases including cancer. Thus, considering the genomic threat posed by DNA damage, cells respond to this DNA damage by activating a complex but distinct network of signalling pathways collectively termed the DNA Damage Response (DDR) that repair damage to constantly maintain the integrity of the genome and prevent the development of diseases such as cancer (Figure 1-8).

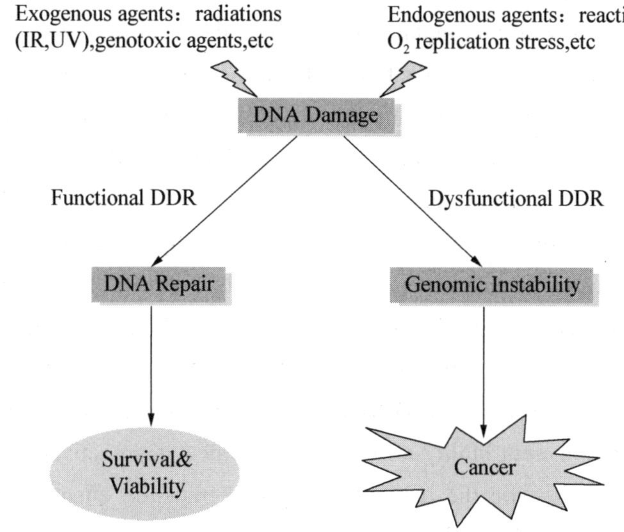

Figure 1-8 DDR and cancer development

The presence of DNA damage either by exogenous or endogenous agents triggers the functional mechanisms of DDR leading to the rapid and efficient repair of DNA damage through cell cycle arrest and delays, and in some cases, apoptosis of cells when DNA damages accumulate beyond repair. This maintains genomic integrity which is critical for cell survival and viability. In contrast, DDR dysfunctions, which may be due to mutations and/or dysregulation of DDR mechanisms, can lead to inefficient or unrepaired DNA damage that in turn destabilize the genome of these cells. Genomic instability induces various aberrant cellular behaviours leading to the development of cancers.

The DDR comprises processes and mechanisms through which DNA damage is detected and repaired to maintain genomic stability and integrity, and this is significant to the survival and viability of a cell or organism. DDR pathways are crucial to both the development of cancers and their treatments, as cancer cells with defective DDR mechanisms exhibit high sensitivity to certain therapeutics, notably DNA damaging agents and also many of these defects are known to drive cancer formation.

Genomic instability is a widely known hallmark of cancer, which may occur as a result of dysfunctional and/or dysregulation of DDR mechanisms. For example, in hereditary cancers such as breast cancer, genomic instability, which drives these cancers, is known to result from mutations in DNA repair genes such as BRCA 1/2. Defects in DDR mechanisms are generally known as major drivers of development in most cancers, such that functional loss (mutation) and/or dysregulation of key DDR genes and proteins are the primary molecular features of many cancers. Such deficiencies in DDR either confer a growth advantage on tumors, thereby breaking the proliferation barrier posed by DDR and allowing the aggressive transformation of pre-cancerous cells to malignant tumors, or they increase the risk of cancer development. For example, women with mutated BRCA genes are more predisposed to developing breast and ovarian cancers than those with non-mutated genes. It is also known, however, and has become more apparent recently that the impairment of DDR mechanisms may significantly impact on the success (or otherwise) of cancer treatments, especially DNA damaging therapies such as cisplatin, Irinotecan, Gemcitabine, and ionizing radiation (IR). These DNA-damaging therapies function by inducing DNA damage, which is cytotoxic to highly proliferating cells. The response of cancer cells to such DNA damage therefore critically determines the success of these treatments. The inability of the cancer cells to efficiently and rapidly repair sufficiently high levels of DNA damage due to their DDR impairments will ultimately lead to cell death and hence account for the increased treatment efficacy of these therapies.

This positive therapeutic impact of DDR impairment of cancer cells on cancer treatment is mostly overcome by the inherent ability of tumors to activate (or re-activate) DDR mechanisms as a strategic response to escape the potentially lethal therapeutic effects of these anti-cancer therapies, however. This perhaps explains the poor response and tumour resistance often observed in most solid tumors with these agents in the clinic. For example, tumour cells that have been shown to emerge as resistant to the DNA-damaging agents (cisplatin and gemcitabine) after prolonged treatments, are also accompanied by extremely high expression of DDR genes and proteins. These findings clearly underscore the crucial role of DDR pathways in both the development and treatment of cancers. DDR and its regulators have therefore become attractive and promising strategic targets for novel cancer therapy: the exploitation of these pathways can provide a platform to develop novel anti-cancer drugs that can act as chemo-and/

or radiosensitizers to enhance the therapeutic response of current conventional DNA-damaging anti-cancer therapies. Moreover, the interplay between major DDR regulators in response to DNA damage creates a 'synthetic lethality' like dependency, where functional loss of DDR components as observed in most tumour cells leads to a greater reliance on the residual DDR factors to maintain viability and survival following DNA damage. Tumour cells are known to exhibit severe and excessive DNA damage during tumorigenesis due to oncogenic-induced replication stress and genomic instability, which may trigger apoptosis or senescence of these cells if left unrepaired. Thus, with most DDR key regulators, including ATM and P53, being mutated or dysregulated in the majority of tumors, tumour cells are more likely to rely on residual pathways such as the ATR pathway in order to repair and survive this self-inflicted excessive DNA damage, and its consequential cell death.

Targeting these residual DDR pathways may therefore be selectively toxic to cancer cells with mutations in certain DDR genes. An example of this 'synthetic lethality' concept is the inhibition of poly ADP-ribose polymerase (PARP), a DNA damage repair enzyme, which has been shown to be selectively lethal to cancer cells harbouring mutations in some DDR genes, including BRCA 1/2. This observation has led to the development of PARP inhibitors olaparib, rucaparib and niraparib which are all currently approved by the Food and Drug Administration (FDA) and European Medicine Agency (EMA) for the treatment of ovarian cancers. Olaparib and rucaparib have also been approved by the FDA for the treatment of prostate cancer (BRCA 1/2 and ATM gene mutated), with various other candidates including veliparib, talazoparib and fluzoparib at various stages of clinical development.

These DDR response signalling pathways have in recent times gained significant attention in cancer therapy, with various DDR proteins including ATR, ATM, DNA-PK, CHK1, CHK2, Wee1, and PARP now all considered promising targets for anti-cancer drug development. With the exception of PARP inhibitors (which have been successfully approved for clinical use) inhibitors of other DDR proteins have yet to realise their clinical potential. The ATR-CHK1 pathway, a major pathway of the DDR machinery, is one of most (if not the most) researched DDR pathways in cancer. Despite over two decades of enormous efforts, an ATR inhibitor for cancer treatment is yet to be clinically successful. In this review, we comprehensively discuss the functional role of ATR in DDR, particularly in relation to cancer development and treatment, and focus on the preclinical studies describing potential ATR inhibition in cancer therapy; hence the rationale behind the development of ATR inhibitors for cancer treatment. The progress and current status of all published ATR inhibitors (13 at the time of writing) since the report of the first (Schizandrin B), including the new generation of ATR inhibitors that has entered clinical development, are also discussed. We conclude by offering some insights into concerns that have been raised and reportedly observed following use of ATR inhibitors in the clinic, and consider how these shortcomings may be addressed so that this promising class of agent fulfils its potential.

引自:BARNIEH F M, LOADMAN P M, FALCONER R A. Progress towards a clinically-successful ATR inhibitor for cancer therapy. Current Research in Pharmacology and Drug Discovery, 2021, 2, 100017.

Words and Expressions

clinically	*adv.* 临床地	radiosensitizers	*n.* 放射增敏剂
integrity	*n.* 完整	metabolism	*n.* 代谢物
endogenous	*adj.* 内源的	lethal	*adj.* 致死的
genomic	*adj.* 基因组的	dysfunction	*n.* 功能失调
mutation	*n.* 体细胞突变	dysregulation	*n.* 异常调控
hallmark	*n.* 检验印记	hereditary	*adj.* 遗传的
tumour	*n.* 肿瘤	cisplatin	*n.* 顺铂
Irinotecan	*n.* 伊立替康	Gemcitabine	*n.* 吉西他滨
synthetic lethality	合成致死	ovarian cancer	卵巢癌
ionizing radiation	电离辐射	prostate cancer	前列腺癌
schizandrin B	五味子素 B	non-mutated gene	非变异基因

Notes

① The DNA Damage Response (DDR)
DNA 损伤应答

② An example of this 'synthetic lethality' concept is the inhibition of poly ADP-ribose polymerase (PARP), a DNA damage repair enzyme, which has been shown to be selectively lethal to cancer cells harbouring mutations in some DDR genes, including BRCA1/2.
"合成致死"概念的一个例子是腺苷二磷酸核糖聚合酶抑制剂,它是 DNA 损伤修复酶抑制剂,这种抑制剂已经被证明对某些 DDR 基因(包括 BRCA1/2)突变的癌细胞具有选择性致死作用。

③ Food and Drug Administration (FDA)、European Medicine Agency (EMA)
美国食品药品监督管理局、欧洲药品管理局

④ ATR(ataxia telangiectasia and Rad3-related)、ATM(ataxia telangiectasia-mutated gene)
共济失调毛细血管扩张突变基因 Rad3 相关激酶、共济失调毛细血管扩张突变

Exercises

1.Put the following into English:

临床　　代谢物　　内源的　　功能失调　　细胞凋亡

2. Put the following into Chinese:

a. Considering the promising preclinical data for these agents, there remains considerable optimism for successful clinical outcomes in these trials.

b. Several dose-limiting toxicities were reported to be observed with AZD6738 dosing, including changes in food consumption and body weight in dogs, rats and mice, in addition to bone marrow toxicity, hypocellularity in multiple lymphoid tissues and increase in alveolar macrophages, although recovery from these toxicities was observed after the termination of dosing.

c. It is worth stating that despite the selectivity, potency and promising pre-clinical data for this generation of ATR inhibitors, the clinical success of these agents particularly in combination with chemotherapy, radiotherapy and immunotherapy will rely on smart dosing strategies that limit the occurrence of haematological and other toxicities.

Reading material:

China's Virus Containment Wins Praise

Unit 8
New Anticancer Agents: Role of Clinical Pharmacy Services

From the pharmacist's point of view and when compared to other clinical disciplines, oncology is characterized by the rapid availability of numerous, costly anticancer agents with new mechanisms of action and sometimes very narrow indications. Based on clinical trials registered between 2007 and 2010 on ClinicalTrials. gov, oncology represents the largest discipline (21.8% of the trials), followed by mental health (9%) and infectious diseases (8.3%). In addition, among the drugs approved in 2012 (39 in the United States and 33 in the European Union), 25%-33% were anticancer agents (13/39 in the United States; 8/33 in the European Union). Another characteristic of oncology has been the development of oral chemotherapy, as well as the re-birth of the subcutaneous route offering patients a more convenient ambulatory management and the possibility of treatment at home. In this context of a fast-growing oncology drug market, both in the community and in hospitals, pharmacists are faced with complex and evoluting information while they have to optimize and secure all these new therapies.

Clinical pharmacy (or clinical pharmacy services) aims to contribute to safe medication use by providing comprehensive management to patients and medical staff, both in the community and the hospital. Pharmacist intervention outcomes include medication appropriateness, adverse drug events, patient satisfaction and economics.

Clinical pharmacy and oncology

Clinical pharmacy, as a discipline, may not be known by all physicians but underlines the evolution of the profession of trained pharmacists from drug distribution and chemotherapy preparation to patient-centered services. Clinical pharmacy in oncology is not very well described; a PubMed search (January 5th, 2014) using the terms "clinical pharmacy services and oncology" only retrieves 229 articles since 1976. In oncology, these services include comprehensive medication reviews integrating chemotherapy, supportive care and ambulatory treatment for co-morbidities, therapeutic drug monitoring (anticancer agents, anti-infective agents, immunosuppressive drugs in recipients of allogeneic stem cell transplantation), supportive care counseling (nutritional support, pain management, chemotherapy side-effects, prophylaxis and treatment), medication information for the medical staff and patients including promotion of adherence to ambulatory treatments, elaboration of therapeutic guidelines, optimal use of economic resources (Table 1-3). Consequently, clinical pharmacists support the multidisciplinary management of patients with cancer.

Table 1-3 Clinical pharmacy services in oncology

Medication review with reconciliation (admission, hospitalization, discharge), integrating anticancer treatment, supportive care, ambulatory treatment, self medication
Drug use optimization integrating genomics and characteristics of the patient
Drug dosing optimization integrating therapeutic drug monitoring (anti-infective agents, anticancer agents, immunosuppressive drugs in recipients of allogeneic stem cell transplantation), pharmacogenomics and body size considerations
Medical staff and patient information and counseling, promotion of adherence to ambulatory treatments
Elaboration of therapeutic guidelines
Optimal use of economic resources

Optimized medication review implies a comprehensive and accurate list of medications taken by the patient (prescription drugs and self-medication). Most cancer patients are aged above 65 years and often have other diseases. In hospitals, the review can be preceded by a phase called reconciliation, generally performed by pharmacists, which aims to identify and correct medication discrepancies during transition care, for example at admission (i.e. to verify that all the drugs taken by the patient at home have been integrated into the medical record by the oncologist). A study performed in the United States found that 24% of the prescription drugs from 152 patients undergoing chemotherapy in the clinic were missing in their medical records. Reconciliation can also be done at discharge because medication regimens might have change during hospitalization. To avoid any re-admission, it is important that patients understand these modifications before returning home (i.e. discontinue drugs that are no longer prescribed, for example). Far beyond the simple detection of drug-drug interactions, medication review focuses on the identification of medication problems (or drug-related problems). Medication problems include inappropriate medications, inappropriate dosing and mode of administration, drug-drug interactions, drug omissions, lack of monitoring. To solve these problems, pharmaceutical interventions, generally, lead to drug dosing adjustments that may be optimized in certain cases by the use of pharmacogenomics and therapeutic drug monitoring, treatment discontinuations, drug additions and replacement of one drug by another. In our experience, in a population of 212 adult hospitalized cancer patients (2, 572 prescriptions including chemotherapy and support), the integration of clinical pharmacy services resulted after medication review in drug-specific interventions for 10% of the prescriptions. Overall, 20% of the interventions concerned inappropriate medications. Drug-drug interactions were reported in 10% of the interventions (representing 1% of the prescriptions). Most of the interventions concerned anti-infective agents and the intervention acceptance rate by oncologists was high (97%). A Dutch study reported a higher rate of medication problems (20%) in a population of 546 patients receiving anticancer treatment. Drug problems mainly concerned contraindications and drug-drug interactions.

Although the intervention of a clinical pharmacist appears beneficial for the cancer patient in the clinic, their value in terms of improved care has not been evaluated in a randomized trial. Research is, therefore, required to better understand the role of clinical pharmacy services in oncology. Further research also focuses on the improvement of pharmaceutical interventions and encompasses many areas of therapeutics and pharmacology. For example, it includes the integration of pharmacogenomics data such as in acute lymphoblastic leukemia where cancer and germline genomics have proved to guide the treatment. It also focuses on the detection and prevention of drug interactions (with concomitant drugs, food and beveredges), benefiting from the development of molecular pharmacokinetics and the availability of updated regulatory guidances for industry (Table 1-4).

Table 1-4 Clinical pharmacy: research in oncology

Impact of clinical pharmacy services in terms of clinical improvement, patient satisfaction and compliance in the clinic and in the community
Evaluation of activity and tolerance of new anticancer agents in real-life studies
Patterns of use of supportive-care drugs (antibacterial agents, antifungal agents)
Development and integration of measures of supportive care
Evaluation and analysis of off-label use of anticancer agents
Analysis of unexpected severe side-effects or low efficacy
Exploration of drug interactions (with food, beveredges, concomitant drugs)
Development and integration of therapeutic drug monitoring and pharmacogenomics in drug use optimization
Impact of body size on pharmacokinetic and clinical variability of anticancer agents
Pharmacoeconomical studies

Clinical pharmacy and new anticancer agents

The research of new anticancer agents follows different approaches such as the continuation of classic cytotoxic drugs and endocrine therapies (eribulin, abiraterone, enzalutamide), the development of analogs and new formulations (vinflunine, cabazitazel, pixantrone, pomalidomide, subcutaneous trastuzumab, liposomal vincristine), and targeted therapy. Most of the new anticancer agents are 'targeted therapies' or, in other words, drugs whose development is based on a pre-determined tumoral or endogenous target. These drugs are either monoclonal antibodies that interact with cell membrane receptors or circulating ligands or protein/enzyme inhibitors that interfere with various tumoral signaling pathways. Monoclonal antibodies are injected intermittently by the intravenous or subcutaneous route, while the protein/kinase inhibitors are mostly administered chronically by the oral route. For clinical pharmacists, these new entities deserve renewed attention. Oral agents are subject to drug-drug interactions both as victims but also as perpetrators in relation to their chronic administration. Furthermore, the development of oral/subcutaneous treatment for ambulatory patients require counseling for optimal compliance. Targeted agents display new mechanisms of action that generate unusual side-effects and novel supportive-care measures, including prevention and the intervention of collaborating specialists (cardiologists, dermatologists). In addition, some agents have narrow indications that are prone to off-label use. In all, clinical pharmacists have a role in the monitoring of new anticancer agents. Among others, we focus on medication problems and pharmaceutical interventions related to oral agents and unusual side-effects (Table 1-5).

Table 1-5 Examples of medication problems and pharmaceutical interventions with new anticancer agents

Oral agents and drug interactions: promote accurate, exhaustive medication review
High risk of drug-drug interactions with kinase inhibitors and abiraterone (as substrates and perpetrators)
High risk of drug-drug interactions with enzalutamide (inducer of CYP3A4, CYP2C9, CYP2C19, CYP1A2, UGT, P-gp, BCRP, OATP1B1) in metastatic castration-resistant prostate cancer
High risk of toxicity when pomalidomide is given with macrolides (CYP3A/P-gp inhibitors) or ciprofloxacin (CYP1A2 inhibitor) in patients with myeloma with bacterial infection
Increased risk of QT prolongation and sudden death when nilotinib is given with food
Decreased oral absorption when erlotinib, gefitinib, dasatinib, sorafenib, bosutinib, vismedogib are administrated with pH modifiers
Unusual side-effects
Skin toxicity with epidermal growth factor receptor antagonists (promote pre-emptive treatment)
Cardiovascular toxicity with antiHER2 agents, antiangiogenic agents, tyrosine kinase inhibitors (promote vigilance, monitoring)
Hypertension, proteinuria, thrombotic events, hemmorhage, gastrointestinal perforation with antiangiogenic agents (promote vigilance, monitoring and contraindications)
Hypothyroidism with sunitinib (promote monitoring)
Hepatotoxicity with the off-label association vemurafenib and ipilimumab in patients with metastatic melanoma (in a general way, promote vigilance with off-label combinations of new agents)

Oral agents. More than 60 oral anticancer agents are now available and among them, 22 kinase inhibitors have been approved worldwide since 2001 (Imatinib). Contrasting with previous oral cytotoxic agents (idarubicin, fludarabine, cyclophosphamide, procarbazine), kinase inhibitors display a higher risk of drug-drug interactions because they are almost both substrates and inhibitors of major pharmacokinetic determinants (cytochrome P450 3A4 or CYP3A4, p-glycoprotein or P-gp) and they are administered chronically. They can generate drug-drug interactions both as substrates and active agents. This also applies to abiraterone, a CYP3A substrate and a CYP2D6, CYP2C8 inhibitor which is used in the treatment of metastatic castration-resistant prostate cancer. Interactions with food, beverages (fruit juices), and pH modifiers may also occur, leading to variations in intestinal absorption. Fortunately, these interactions may be prevented due to better documentation. This information followed new regulatory guidance and is included in the package insert. However, pharmacists should be aware of the risk of interactions with these new oral agents and should provide appropriate information to the oncologist and to the patients. For example, nilotinib must be taken without food because the ingestion of a high fat meal greatly increases the oral absorption (+50%) inpatients with cancer, leading to a risk of prolongation of the electrocardiographic QT interval and sudden death. Overexposure also occurs (ten-fold) when abiraterone acetate (the oral pro-drug of abiraterone) is taken with food. Furthermore, it underlines the debatable strategy of labeling these poorly-absorbable drugs in the fasted state. Pomalidomide is a new oral immunomodulatory agent used in the treatment of refractory myeloma. Contrasting with its analogs lenalidomide and thalidomide, pomalidomide pharmacokinetics are subject to P-gp and CYP3A4/CYP1A2 inhibition. Based on a randomized clinical trial, the use of pomalidomide combined with low-dose dexamethasone was characterized by a significant incidence of febrile neutropenia (10%) and pneumonia (12%). Certain anti-bacterial agents are CYP1A2 inhibitors (ciprofloxacin) or CYP3A/P-gp inhibitors (macrolides except spiramycin). Giving pomalidomide with these antibacterial agents may be contraproductive since it can exacerbate the severity of infection. The oral absorption of some kinase/protein inhibitors (erlotinib, gefitinib, dasatinib, sorafenib, bosutinib, vismedogib) is decreased with concomitant administration of pH modifiers (proton pump inhibitors, H2-receptor antagonists, antacids). Elevated gastric pH causes decreased solubility and impairs absorption. This interaction is of importance given the high prevalence of pH modifiers use among cancer patients. Based on two healthcare databases, a study has estimated that 20%–33% of patients with cancer in the United States received a pH modifier, mostly a proton pump inhibitor. Enzalutamide is a new oral antiandrogen agent which is approved for treatment of metastatic castration-resistant prostate cancer. According to the package insert, enzalutamide is an inducer of many enzymes (CYP3A4, CYP2C9, CYP2C19, CYP1A2, UDP glucuronosyltransferase or UGT) and drug transporters (P-gp, Breast Cancer Resistance Protein or BCRP, Organic Anion Transporting Polypeptide 1B1 or OATP1B1). Enzalutamide has the potential to alter the pharmacokinetics of numerous co-administrated drugs. The risk of drug-drug interactions is huge and necessitates a careful and exhaustive medication review. Enzalutamide is added to the limited list of inducers including rifampin, rifabutin, phenobarbital, carbamazepine, phenytoin, efavirenz, mitotane and bosentan.

Side-effects. When compared to conventional chemotherapy, targeted-therapies differ in their pattern of side-effects. These toxicities, whose origins are related to the mechanisms of action (on/off target) or remain unknown, can impact on the patient's quality of life and lead to treatment discontinuation. Kinase inhibitors are often and erroneously considered as safe medications. On October 31, 2013, in the U.S. the FDA suspended the marketing of ponatinib (initially approved in December 2012) used in the treatment of chronic myelogenous leukemia and Philadelphia-positive acute lymphoblastic leukemia-intolerant or-resistant to prior kinase inhibitor therapy patients, due to life-threatening blood clots and severe narrowing of blood vessels. This severe side-effect occurred in 11.8% of the patients included in a phase Ⅱ trial over a follow-up period of two years. For the first time, a kinase inhibitor was withdrawn from the market for toxicities reasons. The FDA announced on December 20, 2013 that the marketing of ponatinib could be resumed to a narrower population (those with resistance mutation T315I, that is to say 20% of the target mutations or for whom no other kinase inhibitor is appropriate) integrating new safety measures. In a general way, this emphasizes the reinforcement of vigilance for rapidly approved agents.

Drugs interacting with epidermal growth factor receptor (cetuximab, panitumumab, erlotinib, gefitinib, lapatinib, afatinib) induce various dermatological side-effects (acneiform skin rash, xerosis, pruritus, paronychia, hair change) in more than 70% of the patients. These effects can be partially prevented (50% reduction) by the use of tetracyclines (doxycycline), sunscreen, topical steroids, skin moisturizer.

Cardiac events (hypertension, heart failure, left ventricular dysfunction, QT prolongation, thrombotic events) are reported in patients treated by bevacizumab, trastuzumab and some kinase inhibitors (sorafenib, sunitinib, pazopanib, cabozantinib, dasatinib, nilotinib, vandetanib). For example, heart failure or left ventricular dysfunction have been reported in 6% of patients under sunitinib. These cardiac side-effects may be exacerbated when kinase inhibitor agents are given with CYP3A inhibitors and/or food (nilotinib). In collaboration with oncologists and cardiologists, clinical pharmacists have a role in the prevention and the monitoring of side-effects through medication reviews and biological surveillance (check for preventive treatments and conditions that increase the risk of cardiac side-effects, such as other QT prolonging agents or drugs that lower plasma potassium and magnesium). Bevacizumab is a monoclonal antibody that targets vascular endothelial growth factor (VEGF), a circulating pro-angiogenic factor. Bevacizumab is indicated in the treatment of various types of solid cancers and its use is associated with numerous side-effects: hemorrhage, hemoptysis, thrombotic events, hypertension, proteinuria, gastrointestinal perforations, stroke, cardiac events, wound-healing complications. Under real life conditions, these side-effects affected 30% of elderly patients. Moreover, 35.5% of the elderly patients under bevacizumab had a contraindication to the treatment.

Hypothyroidism is a frequent side-effect of sunitinib and has been observed in up to 85% of patients. Monitoring for hypothyroidism is required and may necessitate substitution therapy.

Ipilimumab and vemurafenib are both approved as single agents in the treatment of metastatic melanoma. Based on their modest activity and their different mechanisms of action, their association was evaluated in a phase Ⅰ trial. Unfortunately, the study was closed due to liver side-effects. In a general way and beyond cost considerations (26, 000 euros per month for the association), this must reinforce vigilance in the case of off-label association even if pre-clinical data are favourable.

Conclusion

Clinical pharmacists with oncology training have the potential to optimize the use of new anticancer agents both in the hospital and the community. With the understanding and recognition of drug interactions and side-effects, pharmacists can provide timely interventions and information to health providers, as well as counseling to patients. Further research is required to evaluate their influence in terms of improved care in randomized trials.

引自：LEVEQUE D, DELPEUCH A, GOURIEUX B. New anticancer agents: role of clinical pharmacy services[J]. Anticancer Research, 2014, 34(4): 1573.

Words and Expressions

pharmacist	*n.* 药剂师	appropriateness	*n.* 适宜性
oncology	*n.* 肿瘤学	nutritional	*adj.* 营养的
prescription	*n.* 处方	pharmacogenomics	*n.* 药物基因组学
cytotoxic	*adj.* 细胞毒性的	endocrine	*adj.* 内分泌的
formulation	*n.* 制剂	monoclonal	*adj.* 单克隆的
intravenous	*adj.* 静脉内、注入静脉的	subcutaneous	*adj.* 皮下的
oral	*adj.* 口服的	cardiologist	*n.* 心内科医生
dermatologist	*n.* 皮肤科医师	Imatinib	*n.* 伊马替尼
chronically	*adv.* 慢性地	abiraterone	*n.* 阿比特龙
electrocardiographic	*adj.* 心电图的	pomalidomide	*n.* 泊马度胺
immunomodulatory	*n.* 免疫调节	myeloma	*n.* 骨髓瘤
pneumonia	*n.* 肺炎	antagonist	*n.* 拮抗剂
antacid	*n.* 抗酸剂	healthcare	*n.* 医疗卫生服务
antiandrogen	*n.* 抗雄激素物质	hypothyroidism	*n.* 甲状腺功能减退
medical records	病例档案	package insert	包装说明书
clinical pharmacy	临床药学	clinical trial	临床试验
side-effect	副反应	clinical disciplines	临床学科
co-administrated drugs	联合用药		

Notes

① Most of the new anticancer agents are 'targeted therapies' or, in other words, drugs whose development is based on a pre-determined tumoral or endogenous target.
大多数新的抗癌药物都是靶向治疗,也就是说,其是基于预先确定的肿瘤或内源性靶点开发的药物。

② Some agents have narrow indications that are prone to off-label use.
一些药物的适应证很窄,容易滥用。

③ This also applies to abiraterone, a CYP3A substrate and a CYP2D6, CYP2C8 inhibitor which is used in the treatment of metastatic castration-resistant prostate cancer.
这也适用于阿比特龙。阿比特龙是一种细胞色素 3A 底物和细胞色素 2D6、细胞色素 2C8 抑制剂,用于治疗转移性去势抵抗性前列腺癌。

④ Based on two healthcare databases, a study has estimated that 20%–33% of patients with cancer in the United States received a pH modifier, mostly a proton pump inhibitor.
一项基于两个医疗数据库的研究估计,美国 20%–33%的癌症患者接受了以质子泵抑制剂为主的 pH 调节剂。

Exercises

1. Put the following into English:
肿瘤学　　内分泌　　拮抗剂　　细胞毒性　　临床药学

2. Put the following into Chinese:

a. clinical trials　　chemotherapy　　immunomodulatory　　monoclonal
nutritional　　oral　　pneumonia　　side-effect

b. These toxicities, whose origins are related to the mechanisms of action (on/off target) or remain unknown, can impact on the patient's quality of life and lead to treatment discontinuation.

Reading material:

Substituted 2-Hydrogen-Pyrazole Derivative Serving as Anticancer Drug

Unit 9

Methods for Isolation, Purification and Structural Elucidation of Bioactive Secondary Metabolites from Marine Invertebrates

Introduction

More than 70% of our planet's surface is covered by oceans, and life on Earth has its origin in the sea. In certain marine ecosystems, such as coral reefs or the deep sea floor, experts estimate that the biological diversity is even higher than in tropical rain forests. Many marine invertebrates such as sponges, soft corals or shell-less molluscs are soft-bodied animals that are either sessile or slow moving and usually lack physical defences like protective shells or spines, thus necessitating chemical defences mechanisms such as the ability to synthesize toxic and/or deterrent compounds. These compounds deter predators, keep competitors at bay or paralyze their prey. Examples of fish-deterrent natural products from marine invertebrates include the pyridoacridine alkaloids kuanoniamine C and D from the sponge *Oceanapia* sp., the furanocembranolide 11β, 12β-epoxypukalide produced by Brazilian octocoral *Phyllogorgia dilatata* or the furanosestertepene variabilin of the Caribbean sponge *Ircinia strobilina*. Moreover, bioassay-guided chemical investigation demonstrated that the mollusc *Saccoglossus kowalevskii* was rejected by fishes due to the presence of 2, 3, 4-tribromopyrrole. Many marine-derived compounds show strong biological activities as any natural product released into the water is rapidly diluted and, therefore, needs to be highly potent to exert a significant biological effect. For this reason, and because of the immense biological diversity in the sea as a whole, it is increasingly recognized that a huge number of natural products and novel chemical entities exist in the oceans, with some of them exhibiting biological activities that may also be useful in the quest for finding new drugs with greater efficacy and specificity for the treatment of human diseases as exemplified by the newly admitted marine-derived drugs Prialt and Yondelis.

Marine natural products have attracted the attention of scientists from different disciplines, such as chemistry, pharmacology, biology and ecology. This notion is supported by the fact that, before 1995, about 6, 500 marine natural products had been isolated, whereas this figure has now escalated to more than 19, 000 compounds. The interest in the marine environment has been stimulated by the array of biological activities of marine natural products and hence their potential biomedical applications.

In this protocol, we will give an overview of the methods used for isolation of bioactive metabolites from marine invertebrates that have been successfully employed by our group as well as by other groups in the field as exemplified by several examples taken from our own research that will be referred to in the last section of the protocol.

Sample collection

This is the first step and may be more difficult than when working with terrestrial organisms. This is not only due to difficulties inherent to collection in the marine environment but also due to problems associated with taxonomy and lack of sufficient biological material. This fact is further complicated by the obstacles encountered in the search for adequate conditions for growth and cultivation of marine invertebrates.

The probability of finding useful active metabolites is obviously dependent on the number of screened samples, so the selection of active ones should be based on fast, economic and representative primary tests, e.g., antioxidant (DPPH) and/or cytotoxicity (MTT) assays. To this point, only minute amounts of the biological material have to be consumed, but once isolation of active constituents is initiated, bulk collection of several hundred grams up to 1 kg or even more is usually necessary, and considerable amounts of lyophilized material may be needed to get sufficient quantities of pure compounds for both structural elucidation and bioactivity testing. For this purpose, selective extraction, separation and purification procedures are followed as shown in Figure 1-9. If the pure compound shows interesting biological activity, further pharmacological assays (*in vitro*, *in vivo*, toxicity, tolerated dose and so on) and chemical studies (structure modification, preparation of analogs, structure-activity relationships and so on) have to be carried out to enter the development step of a potential new drug.

Marine organisms can be freshly extracted by means of methanol or ethanol after being collected from their natural habitats or after freeze-drying. In some cases, however, extraction of fresh, sea water-containing material with organic solvents may lead to chemical alterations of compounds due to catalytic conversions of natural products by enzymes that are liberated from their storage compartments during the extraction process25. Thus, lyophilization of the biological material before extraction is considered to be the preferable method.

Critical

1. Generally, all the marine extracts, chromatographic fractions and pure compounds should be kept out of direct sunlight and preferably stored at −20 ℃ as a measure against the potential susceptibility of many marine secondary metabolites to oxidative degradation in air and against isomerization of double bonds in sunlight at room temperature (25 ℃).

2. The time the freshly collected organism is stored in methanol or ethanol should not be unnecessarily prolonged to avoid alkylation or esterification, which gives rise to alkylated artifacts or esters of the secondary metabolites.

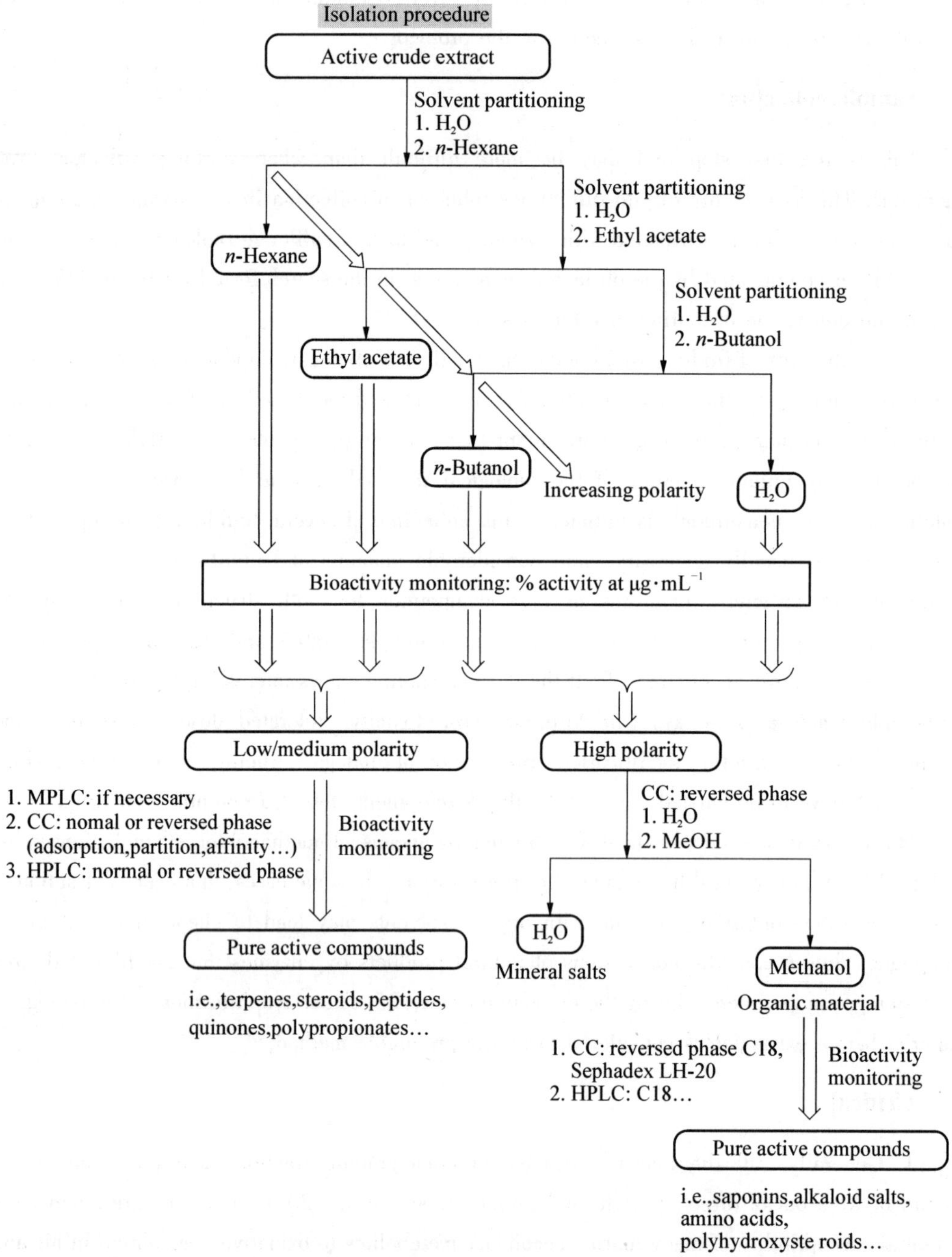

Figure 1-9 Important steps in the search for bioactive constituents from natural sources

Procedure

Isolation procedure of bioactive secondary metabolites from marine invertebrates.

1. Grind the freeze-dried samples and extract for 2-3 cycles, each with 1 L acetone per

100 g of the biomass in order to denaturate cellular proteins (enzymes) and liberate secondary metabolites from the cells. (Note: Each cycle for extraction with acetone should be left overnight at room temperature with stirring either by automatic shaker or magnetic stirrer.)

2. After acetone extraction, extract the remaining residue of the sample with methanol for a number of extraction cycles depending on the weight of the material and allowing enough time to achieve color fading of the biomass and to get optimal extraction of the sample; usually, three cycles are enough. (Note: Each cycle for extraction with methanol stands at room temperature overnight with stirring using either automatic shaker or magnetic stirrer.)

3. Combine the acetone and methanolic extracts and dry under vacuum to give a solid or oily residue. This can be attained by portionwise evaporation of the extract using a rotary evaporator at less than or equal to 40 ℃ till complete solvent evaporation.

4. Dissolve the residue in the smallest possible volume of 10% (by volume) methanol in water and fractionate using separating funnel against n-hexane or petroleum ether (PE) (for defatting) followed by ethyl acetate and n-BuOH.

Critical step: The ratio between the two immiscible solvents should be always kept different (i.e., not 1 : 1) in all liquid-liquid fractionation steps to avoid the formation of an emulsion. In case emulsion has been formed, it can be returned to two immiscible phases by carefully warming the emulsion in a water bath, by addition of salt (NaCl), by centrifugation or by changing the ratio of the solvents. Each step of the solvent fractionation scheme should be carried out with care and should be left to stand till complete separation of the two immiscible liquid phases is achieved.

5. Dry each fraction using a rotavapor to give a solid or an oily residue. All fractions are then subjected to TLC, HPLC, LC/MS and bioactivity assays as well, as described in Step 8.

Critical step: On the basis of the obtained results, evidence on the success of the solvent fractionation can be noticed easily by differences in bioactivities, HPLC profiles and TLC as well as of the different fractions.

6. In accordance with the diverse properties of the components of the fractions, two different procedures for purification can be followed. For low/medium-polarity compounds, refer to option A; for water-soluble compounds, refer to option B.

(A) Low-or medium-polarity fractions

Fractions containing low-or medium-polarity compounds are further fractionated and purified using medium-pressure liquid chromatography, such as VLC or flash chromatographic techniques. Then, purification proceeds further by CC using either normal or reversed stationary phase and a suitable mobile phase to elute the components.

(B) Polar fractions

Highly polar fractions contain water-soluble organic compounds. In our experience, a good procedure is to use reversed-phase CC, eluted gradually from water to MeOH, to eliminate sodium chloride and other mineral salts present in large amounts in these fractions.

7. Continue the purification procedures until you obtain compounds of sufficient purity to allow structural elucidation. This is carried out by using various spectroscopic methods (Figure 1-10), mainly MS and NMR (1D and 2D).

- Structure elucidation
  - Spectroscopic methods
    - MS
      - ESI
      - EI
      - FAB
      - HRMS
    - NMR
      - ^1H, ^{13}C, DEPT
      - COSY, HMBC,
      - HMQC, TOCSY,
      - ROESY, NOESY
  - Stereochemistry—Optical activity, i.e., polarimetry

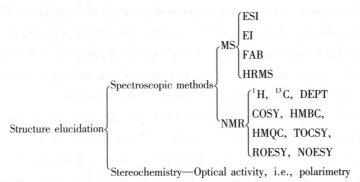

Figure 1-10 Main methods and techniques used in structural elucidation

Critical step: Bioactivity testing as described in the following steps is performed during the whole course to guide the isolation of bioactive secondary metabolites. It starts with the crude extract continuing until purified compounds have been obtained. Then, quantitative assays are required for biological investigation to determine IC_{50} values and to investigate structure-activity relationships using also structurally related components for comparison.

DPPH assay

8. Both qualitative and quantitative antioxidant DPPH assays are performed according to methods reported by Murray et al. The principle is as follows: free radicals, defined as an atom or molecule having at least one unpaired electron, are the main focus in research related to antioxidant and oxidative stress. DPPH is a purple-colored compound that does not dimerize and can hence be prepared in crystalline form. Any molecule that can donate an electron or hydrogen will react with DPPH, thus bleaching its color from a purple-to the yellow-colored diphenylpicryl hydrazine. The qualitative screening (option A) is performed with a rapid TLC screening method using the DPPH radical. The quantitative assay is carried out by performing the steps in option B.

(A) Qualitative screening

a. Perform analytical TLC on precoated TLC plates with silica gel 60 F_{254}. Apply 5 μL of each test extract, fraction or compound solution ($1\ mg \cdot mL^{-1}$).

b. Develop with the appropriate eluent, dry and spray with DPPH solution [0.2% (wt/vol), MeOH].

c. Examine the plates 30 min after spraying. Antioxidant activity is recognized by yellow spots against a purple background. The flavonoids quercetin and luteolin are used as commercially available reference compounds.

(B) Quantitative assay

a. Prepare seven concentrations, ranging from 1 to 100 μM for each sample and analyzed in triplicate. A total of 3.2 mL of MeOH plus 200 μL of each compound solution are used as blank solutions. A total of 3.2 mL of 0.004% (wt/vol) DPPH solution plus 200 μL of MeOH are used as negative control.

b. Add 200 μL of a methanolic solution of the test compound to 3.2 mL of a 0.004% (wt/vol) DPPH solution in MeOH.

c. Determine the absorbance at 517 nm after 30 min of incubation, and the percentage of DPPH reduction is calculated taking into account the absorbance of blank solutions and negative control. Quercetin and/or luteolin are used as reference compounds under the same experimental conditions.

MTT assay

9. Cytotoxicity is tested against L5178Y mouse lymphoma, H4IIE rat hepatoma or C6 rat glioma cell lines using the Cytotoxicity (MTT) assay.

The principle is as follows: MTT, 3-(4, 5-dimethylthiazol-2-yl)-2, 5-diphenyltetra-zolium bromide, is a yellow-colored compound that is converted by mitochondrial reductases into a blue formazan derivative. MTT assay is performed following the method described in 1987 by Carmichael et al., and percentage of cell viability is determined by spectrophotometric determination of accumulated formazan derivative in treated cells at 560 nm in comparison with the untreated ones.

First, we perform the assay qualitatively at a sample concentration of 10 $\mu g \cdot mL^{-1}$. Then, samples that show activity at this concentration are diluted until IC_{50} values, using the following steps, can be determined.

The first step is to prepare stock solutions of the tested samples in EtOH 96% (by volume). All experiments are carried out in triplicate and repeated three times. As negative controls, media with 0.1% (by volume) EtOH are included in the experiments. As positive controls, compounds with known cytotoxicity such as kahalalide derivatives may be used.

10. Harvest exponentially growing cells, count and dilute appropriately. For each sample replicate, pipette 50 μL containing 3, 750 cells into 96-well microtiter plates.

11. Add 50 mL of a solution of the test samples containing the appropriate concentration to each well. We regularly use a concentration range of 3 – 10 $\mu g \cdot mL^{-1}$. For very active compounds, samples may have to be diluted further. The small amount of EtOH present in the wells does not affect the experiments.

12. Incubate the test plates at 37 ℃ with 5% CO_2 for 72 h.

13. Prepare a solution of MTT at 5 $mg \cdot mL^{-1}$ in phosphatebuffered saline (1.5 mM KH_2PO_4, 6.5 mM Na_2HPO_4, 137 mM NaCl, 2.7 mM KCl; pH = 7.4), and from this solution, pipette a volume of 20 μL into each well. The yellow MTT penetrates the healthy living cells, and in the presence of mitochondrial dehydrogenases, MTT is transformed to its blue formazan complex.

14. Incubate the plate for 3 h 45 min at 37 ℃ in a humidified incubator with 5% CO_2.

15. After this incubation, fix the cells on the plate with an aqueous solution containing 1% formaldehyde and 1% (wt/vol) calcium chloride and then lysed with isopropanol: formic acid 95: 5 (vol/vol).

16. Measure the absorbance of the formed formazan product at 520 nm using a scanning microtiter-spectrophotometer. The color intensity is correlated with the number of healthy living cells.

Cell survival is calculated using the formula:

$$\text{Survival } \% = 100 \times \frac{\text{Absorbance of treated cells} - \text{Absorbance of culture medium}}{\text{Absorbance of untreated cells} - \text{Absorbance of culture medium}}$$

Data are given as mean ± s. e. of three independent experiments. The significance of changes in the test responses was assessed using one-way ANOVA; differences were considered significant at $P<0.05$.

17. Study the structure-activity relationships of structurally related compounds to obtain optimized compounds that can be used as drug lead from natural sources.

引自: EBADA S S, EDRADA R A, LIN W H, et al. Methods for isolation, purification and structural elucidation of bioactive secondary metabolites from marine invertebrates [J]. Nature Protocols, 2008, 3(12): 1820–1831.

Words and Expressions

competitor	*n.* 竞争对手,参赛者	Yondelis	*n.* 曲贝替定
paralyze	*vt.* 麻痹;使瘫痪或麻痹	deter	*v.* 阻止,威慑
deterrent	*n.* 威慑因素;*adj.* 遏制的,制止的	lyophilize	*vt.* 使冻干
escalate	*v.* (使)逐步扩大,不断恶化	immense	*adj.* 极大的,巨大的
biomedical	*adj.* 生物医学的	obstacle	*n.* 障碍,阻碍
taxonomy	*n.* 分类学,分类系统	Prialt	*n.* 齐考诺肽
terrestrial	*adj.* 陆地的,路栖的;*n.* 陆地生物	cytotoxicity	*n.* 细胞毒性
preferable	*adj.* 更可取的,较适合的	elucidation	*n.* 说明,阐明
penetrate	*vt.* 穿透,渗透	Quercetin	*n.* 槲皮素
denature	*vt.* 使变质,使中毒	predator	*n.* 捕食者,掠夺者
fractionate	*vt.* 分馏,把……分成几部分	glioma	*n.* 神经胶细胞瘤
immiscible	*adj.* 不混溶的,不互溶的	absorbance	*n.* 吸光度,吸光率

reduction	n. 还原,减少	incubation	n. 孵化;(传染病的)潜伏期;
reference	n. 参考,参照	biomass	n. 生物量(以生境面积等表示)
hepatoma	n. 肝癌	viability	n. 生命力,耐用性,寿命
reductase	n. 还原酶	lyophilization	n. 冻干法
isopropanol	n. 异丙醇	susceptibility	n. 敏感性,易受影响的特性
dehydrogenases	脱氢酶	bioactivity testing	生物活性测试
magnetic stirrer	磁力搅拌器	flash chromatography	快速色谱法(柱层析)
automatic shaker	自动摇动器		

Notes

① *n*-hexane 和 *n*-BuOH。其中 *n* 原型是 normal,常缩写为 *n*,在有机化合物名称中表示"正"。-ane 表示烷烃的词尾;hexa 表示数词"六",或有机化合物中六个碳原子"己",故 *n*-hexane 译为正己烷;Bu 表示词头 buta-,意指"四、丁",*n*-BuOH 中 OH 表示"醇羟基",故译为正丁醇。

② TLC 和 HPLC 是专业英语构词法中常见的一种方法,叫作压缩法或缩略法。具体见第六章第四单元。TLC 原型是 thin layer chromatography,译为薄层色谱法,它是将适宜的固定相涂布于玻璃板、塑料或铝基片上,成一均匀薄层。待点样、展开后,根据比移值(R_f)与适宜的对照物按同法所得的色谱图的比移值(R_f)做对比,用以进行药品的鉴别、杂质检查或含量测定的方法。HPLC 原型是 high performance liquid chromatography,译为高效液相色谱法,以液体为流动相,采用高压输液系统,将具有不同极性的单一溶剂或不同比例的混合溶剂、缓冲液等流动相泵入装有固定相的色谱柱,在柱内各成分被分离后,进入检测器进行检测,从而实现对试样的分析。

③ Continue the purification procedures until you obtain compounds of sufficient purity to allow structural elucidation. This is carried out by using various spectroscopic methods, mainly MS and NMR (1D and 2D).

首先,第一句翻译时需要根据翻译技巧——减词法原则"英语中有些代词在译成汉语时必须省略,以使译文简洁,句意明白无误",不翻译出句子中的单词"you"。如:We cannot see sound wave as they travel through air. 不需要翻译 we,译为"声波在空气中传播时,是看不见的"。该翻译技巧在专业英语中比较常见。此外,第二句中是被动句,被动句在专业英语中很常见,约占 1/3。此处应该采用翻译为主动句的翻译技巧。被动句翻译技巧具体见第六章。

④ If the pure compound shows interesting biological activity, further pharmacological assays (*in vitro*, *in vivo*, toxicity, tolerated dose and so on) and chemical studies (structure modification, preparation of analogs, structure-activity relationships and so on) have to be carried out to enter the development step of a potential new drug.

在专业文献中,interesting 经常出现在 introduction 部分,表示“吸引人的,引人关注的”。new drug 本句中译为新药,通常 drug 含义比较广泛,指任何用于预防或治疗肉体上或精神上疾病的药品,尤其多用来表示“毒品”,而 medicine 指医生开给病人用的一切药物,表示“药物,医疗”。在汉译英或英译汉时,通常会采用括号、破折号等方法翻译一些解释说明类的文字,比如,本句中 chemical studies (structure modification, preparation of analogs, structure-activity relationships and so on),用括号里面内容进一步解释说明化学研究包括结构修饰、类似物制备、构效关系等。

⑤ For this reason, and because of the immense biological diversity in the sea as a whole, **it** is increasingly recognized **that** a huge number of natural products and novel chemical entities exist in the oceans, with some of them exhibiting biological activities **that** may also be useful in the quest for finding new drugs with greater efficacy and specificity for the treatment of human diseases as exemplified by the newly admitted marine-derived drugs Prialt and Yondelis.

此句为典型复杂长句,是专业英语学习难点之一,需要通过分析句子成分,把复杂长句分解为几个简单句,化繁为简。it is…that…是本句主句的骨架。前面 for this reason, and because of…为原因状语。with some of…that…为介词短语引导的一个复杂从句,其中 that 引导一个定语从句。

Exercises

1. Put the following into English:

阴性对照	冻干法	空白溶液	中药
海洋药物	刺激代谢产物	真空干燥	药理实验

2. Put the following into Chinese:

a.

marine invertebrates	freeze-drying	biological activities	Prialt
biological diversity	cytotoxicity	bioactive metabolites	fractionate

b. If the pure compound shows interesting biological activity, further pharmacological assays (*in vitro*, *in vivo*, toxicity, tolerated dose and so on) and chemical studies (structure modification, preparation of analogs, structure-activity relationships and so on) have to be carried out to enter the development step of a potential new drug.

Reading material A:

Guan Huashi—Pioneers and Founder of Marine Pharmaceuticals in China

Guan Huashi was born in Xiajin, Shandong province, and graduated from Shandong

Maritime College (renamed Ocean University of China) in 1964. He was elected as an academician of the Chinese Academy of Engineering in 1995. As a marine pharmacologist, professor Guan Huashi is one of the most prominent pioneers and founders of Marine Pharmaceuticals in China.

"As a cradle of life, the ocean abounds in pharmacological potential, " said Guan, "Products developed from marine sources can be of therapeutic value." During his childhood, Guan loved to raise aquatic creatures such as fish, shrimps, crabs, turtles and tadpoles. "Maybe my curiosity for creatures in water and my respect for rivers, lakes and seas when I was a child had some bearing on my career, " said Guan. He is in constant pursuit of cures for human diseases, and his source is the ocean.

Professor Guan has been engaged in the development of marine biological resources as well as the teaching and research of marine pharmaceuticals and food engineering. In 1960s, he participated in the project of the Industrialization of Algae Iodine-Extracting Technology. In 1970s, he led and finished the project of Recycled Use of the Byproducts of Kelp after Iodine-extraction, and successfully invented four new products including Emulsifier for Agricultural Use, which hence made substantial contribution to the industry of iodine of China.

Guan had for the first time led a team that successfully developed the country's first marine drug, called PSS, in 1985, which not only greatly benefited society and economy, but also initiated and promoted the development of marine drug research in China. The drug helps to treat cardiovascular diseases, those related to heart or blood vessels, and cerebrovascular diseases, which are related to blood vessels supplying the brain. In 1990s, a series of marine drugs and functional products including Gan-Tang-Zhi, Hai-Li-Te and Jiang-Tang-Ning were successfully invented and put into production. These inventions made fundamental contribution to the rise, development and maturity of marine pharmaceutical industry in China.

In order to promote sustained and healthy development of China's algae industry and further development of glycobiology research, Professor Guan and his team conquered several key problems and made break-through in the use and preparation of marine oligosaccharides including alginate, agar, carrageenan and chitosan. The system of industrialization of the preparation of marine oligosaccharides and the first databank of marine oligosaccharides were built up. Based on the marine oligosaccharide database, four first-class marine drug candidates are currently being studied in different clinical trials and their potential market effect is highly attractive. For example, GV-971, an Alzheimer's drug, is extracted from brown algae. The first phase of clinical trials has been completed. The drug is aimed at patients with mild to moderate Alzheimer's, a chronic neurodegenerative disease that usually starts slowly and gradually worsens over time, destroying thinking skills and the ability to carry out simple tasks. Independent animal experiments found that the drug can regulate the immune system, reduce neuroinflammation and improve cognition. This achievement not only promoted the sustained development of algae industry and further study of marine glycochemistry and glycobiology, but also provided vital theoretical and technological support to the research of marine drugs and new economic growth point.

In 2009, Professor Guan has compiled a comprehensive encyclopedia entitled *Chinese Marine Materia Medica* that is set to become a classic. The encyclopedia of *Chinese Marine Materia Medica* provides an important scientific foundation for the research and development of modern marine drugs and an informative basis for policy makers to make scientific decisions on the use and protection of marine medicinal bioresources.

Guan's team has collaborated with dozens of research groups and more than 100 scientists from five countries. In Guan's view, commercialization is a big step in turning his research findings into practical solutions. Therefore, he and his team maintain close ties with pharmaceutical companies. Guan's team signed an agreement with CP Pharmaceutical Group Qingdao for accelerating transformation of marine pharmaceutical research into commercial products.

Reading material B:

The Odyssey of Marine Pharmaceuticals: A Current Pipeline Perspective

Unit 10

Growth Years and Post-Harvest Processing Methods Have Critical Roles on the Contents of Medicinal Active Ingredients of Scutellaria Baicalensis

Medicinal plants with effective pharmacological activities, low side effects as well as high economic values have long been used as raw materials and reservoirs of new drugs in pharmaceutical industry. As an important part of the traditional Chinese medicine (TCM) system, medicinal plants have shown great therapeutic potentials in severe and acute diseases including malaria, diabetes, cancer, etc. Specifically, in the recent pandemic of corona virus disease 2019 (COVID-19), TCM preparation and drugs showed effective anti-viral and anti-inflammatory activities against the severe acute respiratory syndrome coronavirus 2 (SARS-CoV-2). Exploring the potential pharmaco-logical functions and promoting the introduction and development of TCM will bring benefits to human health globally.

Currently, the primary aim of studies in TCM cultivation and production is how to improve the yield and quality of medicinal crops (specifically the active ingredients with high medicinal values). In recent years, the market demand of raw materials originated from medicinal crops is increasing with the development of pharmaceutical industry. Nevertheless, the yield of medicinal crops with high quality is decreasing due to growing human activities and habitat loss. A pressing challenge faced by scientists is how to obtain the maximum yield of active ingredients in medicinal crops.

Improving the yield of active ingredients by optimizing the processing methods is a good way to achieve the above goals. Typically, the active ingredients in medicinal plants are secondary metabolites that are closely related to cultivation methods, field management, harvest time, primary processing in the producing area, deep processing, storage, and transportation, etc. Specifically, the processing methods including removing impurities, stacking, steaming, slicing, and drying have significant effects on the contents of active ingredients of Chinese medicinal materials. The primary processing in the producing area is the first key step for the quality formation of Chinese medicinal materials. Previous studies have demonstrated that primary processing in the producing area have a great influence on the yield, quality, and efficacy of the medicinal materials. In the traditional technology of primary processing, it is specified that post-harvest treatment and processing of medicinal materials in the producing area should follow the natural rules of the formation of active ingredients. Meanwhile, the specific characteristics of raw materials during the process of storage and transportation should be carefully considered to ensure the quality of medicinal materials.

Nevertheless, most of the active ingredients (e.g., alkaloid salts, inorganic salts, etc.) of TCM have different affinity degrees with water. Improper processing treatments (long-term exposure to water, water flushing, immersion, etc.) will cause considerable loss of active ingredients. In addition, the contents of active ingredients in medicinal materials are closely related to the enzyme activity of plant tissues. For medicinal crops that are rich in alkaloids and polysaccharides, the contents of active ingredients as well as the quality of medicinal materials will rapidly decline, due to the absence of high-temperature treatment that can inactivate enzyme activity in tissues and organs after harvest. For example, for the important medicinal crop tall gastrodia tuber (*Gastrodia elata* Bl.), high-temperature steam is firstly used to inactivate enzyme activity in the root after harvesting, and then the steamed roots are dried. As an alternative, it can also inactivate enzyme activity by boiling the root in water. However, the boiling process will significantly reduce the content of gastrodin, which will further reduce the quality of medicinal materials. Accordingly, it is necessary to investigate the effects of processing methods on the quality formation of different active ingredients in medicinal crops, thereby providing optimum processing methods that will obtain the maximum yield of products for the industrial production of traditional Chinese medicinal materials.

Baikal Skullcap (*Scutellaria baicalensis Georgi*) is a perennial herb in the *Lamiaceae* family and its dried root is used as a famous traditional Chinese medicine, which was first recorded in the *Shennong Bencao Jing* (*The Classic of Herbal Medicine*) for the treatment of diarrhea, dysentery, hypertension, hemorrhaging, insomnia, inflammation, and respiratory infection. As one of the most commonly used Chinese medicinal materials in China, it has been used as medicinal materials for more than 2, 000 years since being recorded in the ancient medical books. Up to date, it has been included in over 90% TCM formulas for treating cold. Modern pharmacological studies show that the active ingredients of *S. baicalensis* have important pharmacological effects such as anti-oxidation, anti-bacterial, anti-viral, anti-tumor, and anti-inflammation. Specifically, it is recently found that *S. baicalensis* has significant curative effects on the treatment of COVID-19. The main pharmaceutical active ingredients are flavonoids, of which the glycosides and the aglycones have the strongest medicinal activities. Among them, the content of baicalin is determined as the main evaluation index and quality control of *S. baicalensis* by the *Chinese Pharmacopoeia*.

Due to its great medicinal values, there is a huge market demand for *S. baicalensis* every year and it is becoming one of the top ten sales of Chinese medicinal materials in China. Before the 1990s, the products of *S. baicalensis* are mainly originated from wild resources. Since the 21 st century, the decreasing wild resources of *S. baicalensis* were far from meeting the rapidly increasing market demand of pharmaceutical use. Accordingly, the market share of cultivated varieties has been increasing after extensive cultivation of *S. baicalensis*.

For the industrial production of *S. baicalensis*, the harvesting time and processing in producing areas are considered as the key factors for the quality formation of medicinal materials. Since *S. baicalensis* is a perennial herb, its stele of the root gradually decays from the base after growing for about 4 years (known as "hollow roots") (Figure 1-11). Although it has been recorded in the classical prescriptions and the medical classics that the hollow roots of *S. baicalensis* has the best curative effect in TCM, the theoretical basis of this statement remains unclear. Meanwhile, the factor of harvesting time is very important for the quality and yield of medicinal materials as well as the income of farmers during the cultivation process. At present, the farmers and herbalists believe that harvesting time at two and a half years or three years can ensure the best quality, the highest yield, and the highest land use efficiency for *S. baicalensis*. However, there is no clear evidence to show whether the contents of medicinal active ingredients of *S. baicalensis* harvested at such time have the highest levels. In addition, the primary processing in producing areas (removing impurities, drying, slicing, etc.) is considered as another key factor that affects the contents of medicinal active ingredients. Previous studies have shown that the contents of medicinal active ingredients in *S. baicalensis* after harvesting are closely related to enzyme activities in roots. The active ingredients such as baicalin and wogonoside in fresh roots are easily reduced in enzyme solution under humid environment or washing and soaking in water. Nevertheless, there is a lack of systematic research on the effects of post-harvest treatment and processing technology on the contents of medicinal active ingredients to reveal such mechanisms. Taken together, it is of great theoretical significance and practical prospect to reveal the effects of different processing technology on the medicinal active ingredients under the background of increasing market demand for *S. baicalensis*.

A. The plant of *S. baicalensis*. From left to right: whole plant, leaf, stem, flower, and root. B. The roots of *S. baicalensis* with different growth years. From left to right: 0.5, 1, 2, 3, 4, 5, 7, 9, 12, 15. C. The different parts of roots from 2-year-old *S. baicalensis*. From left to right: whole root, root sclice, cortex, stele.

Figure 1-11 The plant and main medicinal material of *S. baicalensis* used in pharmaceutical industry

A comprehensive study was conducted from the following aspects by taking eight medicinal active ingredients from the root of *S. baicalensis* as indicators: (1) the contents of active ingredients in roots of *S. baicalensis* from growing 1−15 years were compared and analyzed to determine the optimal harvest time; (2) six medicinal active ingredients in the parts of stele and cortex in roots of 2-year and 15-year old *S. baicalensis* were compared and analyzed to reveal the differences of active ingredients in different growth years; (3) the dynamic changes of the contents of active ingredients in fresh-crush and fresh-cut roots of *S. baicalensis* at room temperature were compared and analyzed to reveal the influence of post-harvest treatment on the contents of active ingredients; (4) the effects of six different post-harvest treatment on the contents of active ingredients were systematically designed and compared to determine the best primary processing technology. Collectively, the above results will provide a theoretical basis for the scientific harvest and the primary processing in the producing area of *S. baicalensis*, and provide a reference for the selection of optimal processing technology of medicinal materials for the industrial production of medicinal crops.

By estimating the contents of eight active ingredients (baicalin, wogonoside, baicalein, wogonin, scutellarin, scutellarein, apigenin, and chrysin) in roots of *S. baicalensis* of different growth years (ranging from 1 year to 15 years) using HPLC, a systematic analysis was conducted to determine the optimal harvest period of this important medicinal crop. The results showed that the best harvest period for *S. baicalensis* should be determined as 2−3 years based on comprehensive evaluation of active ingredient content, annual yield increment, and land use efficiency. Meanwhile, the contents of six active ingredients were compared in different parts (cortex and stele) of roots of *S. baicalensis*, and the dynamic changes of the contents of active ingredients was monitored in fresh-crush and fresh-cut roots of *S. baicalensis* at room temperature. In addition, we observed the effects of six different post-harvest treatments on the contents of active ingredients to determine the best primary processing technology. It can be concluded that the contents of baicalin, wogonoside, baicalein, and wogonin in cortex were significantly higher than those in stele ($P \leqslant 0.05$). The contents of four main active ingredients (baicalin, wogonoside, baicalein, and wogonin) under drying (D) and cutting-drying (C-D) treatments were significantly higher than those of the other four treatments ($P \leqslant 0.05$). Collectively, these results will provide valuable references for determining the optimal harvesting and processing technology that will obtain the maximum products of active ingredients for *S. baicalensis*. Furthermore, the processing methods of softening, cutting, and drying proposed in this study will help improving the product yield for pharmaceutical industry of other medicinal crops.

引自：BAI C K, YANG J J, CAO B, et al. Growth years and post-harvest processing methods have critical roles on the contents of medicinal active ingredients of *Scutellaria baicalensis*[J]. Industrial Crops and Products, 2020, 158, 112985.

Words and Expressions

pharmacological *adj.* 药理学的,药物学的
post-harvest *n.* 收割期后,收后,采后
cultivation *n.* 种植,栽培
stacking *n.* 堆垛; *vt.* 堆积,将……码放整齐
efficacy *n.* 功效;(尤指药物或治疗方法的)效力
slicing *n.* 切割; *vt.* 把……切成(薄)片
anti-viral *n.* 抗病毒的
indicator *n.* 指示信号,(化)指示剂
harvest *n.* 收割,收获; *vt.* 收割(庄稼等)
yield *n.* 收率,产率
perennial *n.* 多年生植物;*adj.* 多年生的,长久的
diabetes *n.*糖尿病
habitat *n.* 栖息地,(动植物的)生活环境
reservoir *n.* 水库,(大量的)储备,储藏
anti-inflammatory *adj.* & *n.* 消炎的,抗炎的
raw material *n.* 原料,原材料
pandemic *n.*(全国或全球性)流行病; *adj.*(疾病)大流行的

pharmaceutical industry 制药工业,医药行业
side effect (药物的)副作用,意外的连带后果
traditional Chinese medicine/TCM 中华医药,中成药,中草药,中医,中药
corona virus disease 2019/COVID-19 2019 冠状病毒病,新型冠状病毒肺炎(简称新冠肺炎)
severe acute respiratory syndrome coronavirus/SARS-CoV 严重急性呼吸综合征冠状病毒
secondary metabolites 次级代谢产物,次生代谢物
respiratory infection 呼吸道感染,呼吸系统感染

Notes

① *S. baicalensis Georgi.* 黄芩,中药名,俗名香水水草、黄筋子,别名山茶根。黄芩属,学名Scutellaria Linn.,是唇形科下的一个属,约 300 余种,广布于世界,但热带非洲少见,非洲南部全无,我国约 100 余种,南北均产之,其中黄芩为著名的中药之一,产于黑龙江、辽宁、内蒙古、四川等地,中国北方多数省区都可种植。其以根入药,味苦、性寒,有清热燥湿、泻火解毒、止血、安胎等功效。主治温热病、上呼吸道感染、肺热咳嗽、湿热黄疸、肺炎、痢疾、咳血、目赤、胎动不安、高血压、痈肿疖疮等症。黄芩的临床抗菌性比黄连好,而且不产生抗药性。

② *Shennong Bencao Jing* (*The Classic of Herbal Medicine*),《神农本草经》,又名《本草经》《本经》,相传起源于神农氏,代代口耳相传,于东汉时期集结整理成书,是中医四大经典著作之一,是已知最早的中药学著作。书中记载了365种药物的疗效,多数真实可靠,仍是临床常用药;并提出了辨证用药的思想,所论药物适应病症能达170多种,对用药剂量、时间等都有具体规定,这也对中药学起到了奠基作用。

③ COVID-19——新型冠状病毒肺炎(Corona Virus Disease 2019),简称为新冠肺炎,世界卫生组织将其命名为2019冠状病毒病,意指2019新型冠状病毒感染导致的肺炎。2019年12月以来,湖北省武汉市部分医院陆续发现了多例有华南海鲜市场暴露史的不明原因肺炎病例,证实为2019新型冠状病毒感染引起的急性呼吸道传染病。2020年2月11日,世界卫生组织总干事谭德塞在瑞士日内瓦宣布,将新型冠状病毒感染的肺炎命名为"COVID-19"。

Exercises

1. Put the following into English:

活性成分	原料	酶活性	中药材
加工方法	抗病毒	抗氧化	黄酮

2. Put the following into Chinese:

TCM	COVID-19	quality control
Inflammation	respiratory infection	anti-inflammation
Pharmacopoeia	the medical classics	post-harvest

Reading material:

Brief Introduction of Cultural Achievements on Traditional Chinese Medicine—*Huangdi Neijing*

The Chinese nation has a long history and a glorious culture, and it is of importance to introduce the nation's greatest cultural achievements to people all over the world.

Ancient China gave birth to a large number of eminent scientists, such as Zu Chongzhi, Li Shizhen, Sun Simiao, Zhang Heng, Shen Kuo and Bi Sheng. They produced numerous treatises on scientific subjects, including *Compendium of Materia Media*, *A Treatise on Febrile Diseases* and *The Manual of Important Arts for the People's Welfare*.

Among the extant ancient medical literatures, the following can be considered as the "classics" of Chinese medicine: *Huangdi Neijing* (*Innar Canon of Yellow Emperor*); *Shennong Bencao Jing* (*Shennong's Herbal, The Great Herbal*); *Shanghan Lun* (*Treatise on Febrile Disease Caused by Cold*); *Zhenjiu Jiayi Jing* (*Acupuncture Classic Jiayi Jing of the Yellow Emperor*), etc.

All the classics, with the exception of *Shengnong's Herbal*, are related to *Huangdi Neijing*, whether it be a research or an elaboration of *Neijing*. From this, we can clearly see the importance of *Neijing*.

Neijing, magnum opus of Chinese medicine, has been taken by scholars of all time as the origin and source of Chinese medicine. But the sphere *Neijing* deals with is not confined with medicine only. In its exploration of the theoretical system concerning the art of healing, it absorbed the then important achievements of a number of subjects including astronomy, calendar and mathematics, meteorology, biology, geography, anthropology, psychology, logic, philosophy, etc. This, on one hand, reflects the inexact classification of scientific studies at that time, but on the other hand, it also reflects the richness *Neijing* embraces. It is a colossal scientific literature radiating great splendor in the history of science of civilization not only in China but also in the world.

As we all know, in the analysis and understanding of human body and diseases, Chinese medicine has a number of unique diagnostic and therapeutic methods and practice with no parallels in Western medicine. The theoretical system of Chinese medicine, composed of theories of orbisconography, Channels and Collaterals, pathology, diagnosis and treatment based on overall analysis of symptoms and signs and acupuncture, has been basically established in the course of the writing of *Neijing*. It is surprising to find out that many medical principles, physiological and pathological phenomena and therapeutics as explored in *Neijing*, a book written 2, 000 years ago, are still active in directing our clinical practice and research, and some of them yet cannot be well explained by modern science, including Western medicine.

The universally acknowledged acupuncture anaesthesia is a new push in world medicine developed on the basis of serious study of the acupuncture and Channels and Collaterals theory of *Neijing*. More and more new discoveries will be made on such studies. Prolongation of life (longevity), Qigong (breathing exercises), meteoropathology, environmental medicine, biochronometry, bionics, treatment based on an organic conception of the human body and phytotherapy are only a few examples of promising future that may be developed in the course of study of *Neijing*.

The art of healing, one of the most important and long-lasting arts in the world, is an art cherished by all people, professional and non-professional, an art with the largest followers and amateurs, an art that none of us can do without and an art that has seen a persistent and continuous development and amelioration in the whole process of human's recorded history of several thousand years. The world medicine, or the modern medicine, is consisted of medical progresses from all countries.

Today, it is impossible for any nation's culture to develop without absorbing the excellent aspects of the cultures of other peoples. When Western culture absorbs aspects of Chinese culture, this is not just because it has come into contact with Chinese culture, but also because of the active creativity and development of Western culture itself; and vice versa. The various cultures of the world's peoples are a precious heritage which we all share. Mankind no longer lives on different continents, but on one big continent, or in a "global village". And so, in this era characterized by an all-encompassing network of knowledge and information, we should learn from each other and march in step along the highway of development to construct a brand-new "global village".

Chapter 2
Drugs Declaration

Unit 1
Impurities in New Drug Substances

1. Preamble

This document is not intended to apply to new drug substances used during the clinical research stage of development. The following types of drug substances are not covered in this document: biological/biotechnological, peptide, oligonucleotide, radiopharmaceutical, fermentation product and semi-synthetic products derived therefrom, herbal products, and crude products of animal or plant origin.

Impurities in new drug substances are addressed from two perspectives:

* *Chemistry Aspects* include classification and identification of impurities, report generation, listing of impurities in specifications, and a brief discussion of analytical procedures.

* *Safety Aspects* include specific guidance for qualifying those impurities that were not present, or were present at substantially lower levels, in batches of a new drug substance used in safety and clinical studies.

2. Classification of impurities

Impurities can be classified into the following categories:

- Organic impurities (process- and drug-related)
- Inorganic impurities
- Residual solvents

Organic impurities can arise during the manufacturing process and/or storage of the new drug substance. They can be identified or unidentified, volatile or non-volatile, and include:

- Starting materials
- By-products
- Intermediates
- Degradation products
- Reagents, ligands and catalysts

Inorganic impurities can result from the manufacturing process. They are normally known and identified and include:

- Reagents, ligands and catalysts

- Heavy metals or other residual metals
- Inorganic salts
- Other materials (e.g., filter aids, charcoal)

Solvents are inorganic or organic liquids used as vehicles for the preparation of solutions or suspensions in the synthesis of a new drug substance. Since these are generally of known toxicity, the selection of appropriate controls is easily accomplished.

Excluded from this document are: (1) extraneous contaminants that should not occur in new drug substances and are more appropriately addressed as Good Manufacturing Practice (GMP) issues, (2) polymorphic forms, and (3) enantiomeric impurities.

3. Rationale for the reporting and control of impurities

3.1 Organic impurities

The applicant should summarise the actual and potential impurities most likely to arise during the synthesis, purification, and storage of the new drug substance. This summary should be based on sound scientific appraisal of the chemical reactions involved in the synthesis, impurities associated with raw materials that could contribute to the impurity profile of the new drug substance, and possible degradation products. This discussion can be limited to those impurities that might reasonably be expected based on knowledge of the chemical reactions and conditions involved.

In addition, the applicant should summarise the laboratory studies conducted to detect impurities in the new drug substance. This summary should include test results of batches manufactured during the development process and batches from the proposed commercial process, as well as the results of stress testing used to identify potential impurities arising during storage. The impurity profile of the drug substance batches intended for marketing should be compared with those used in development, and any differences discussed.

The studies conducted to characterise the structure of actual impurities present in the new drug substance at a level greater than (>) the identification threshold given in Table 2-1 (e.g., calculated using the response factor of the drug substance) should be described. Note that any impurity at a level greater than (>) the identification threshold in any batch manufactured by the proposed commercial process should be identified. In addition, any degradation product observed in stability studies at recommended storage conditions at a level greater than (>) the identification threshold should be identified. When identification of an impurity is not feasible, a summary of the laboratory studies demonstrating the unsuccessful effort should be included in the application. Where attempts have been made to identify impurities present at levels of not more than (≤) the identification thresholds, it is useful also to report the results of these studies.

Table 2-1 Thresholds

Maximum Daily Dose[1]	Reporting Threshold[2,3]	Identification Threshold[3]	Qualification Threshold[3]
≤ 2 g/day	0.05%	0.10% or 1.0 mg per day intake (whichever is lower)	0.15% or 1.0 mg per day intake (whichever is lower)
> 2 g/day	0.03%	0.05%	0.05%

[1] The amount of drug substance administered per day

[2] Higher reporting thresholds should be scientifically justified

[3] Lower thresholds can be appropriate if the impurity is unusually toxic

Identification of impurities present at an apparent level of not more than (≤) the identification threshold is generally not considered necessary. However, analytical procedures should be developed for those potential impurities that are expected to be unusually potent, producing toxic or pharmacological effects at a level not more than (≤) the identification threshold. All impurities should be qualified as described later in ICH harmonized tripartite guideline.

3.2 Inorganic impurities

Inorganic impurities are normally detected and quantified using pharmacopoeial or other appropriate procedures. Carry-over of catalysts to the new drug substance should be evaluated during development. The need for inclusion or exclusion of inorganic impurities in the new drug substance specification should be discussed. Acceptance criteria should be based on pharmacopoeial standards or known safety data.

3.3 Solvents

The control of residues of the solvents used in the manufacturing process for the new drug substance should be discussed and presented according to the ICH Q3C Guideline for Residual Solvents.

4. Listing of impurities in specifications

The specification for a new drug substance should include a list of impurities. Stability studies, chemical development studies, and routine batch analyses can be used to predict those impurities likely to occur in the commercial product. The selection of impurities in the new drug substance specification should be based on the impurities found in batches manufactured by the proposed commercial process. Those individual impurities with specific acceptance criteria included in the specification for the new drug substance are referred to as "specified impurities" in this guideline. Specified impurities can be identified or unidentified.

A rationale for the inclusion or exclusion of impurities in the specification should be presented. This rationale should include a discussion of the impurity profiles observed in the safety and clinical development batches, together with a consideration of the impurity profile of batches manufactured by the proposed commercial process. Specified identified impurities should be included along with specified unidentified impurities estimated to be present at a level greater than (>) the identification threshold given in Table 2-1. For impurities known to be unusually potent or to produce toxic or unexpected pharmacological effects, the quantitation/detection limit of the analytical procedures should be commensurate with the level at which the impurities should be controlled. For unidentified impurities, the procedure used and assumptions made in establishing the level of the impurity should be clearly stated. Specified, unidentified impurities should be referred to by an appropriate qualitative analytical descriptive label (e.g., "unidentified A", "unidentified with relative retention of 0.9"). A general acceptance criterion of not more than (≤) the identification threshold (Table 2-1) for any unspecified impurity and an acceptance criterion for total impurities should be included.

Acceptance criteria should be set no higher than the level that can be justified by safety data, and should be consistent with the level achievable by the manufacturing process and the analytical capability. Where there is no safety concern, impurity acceptance criteria should be based on data generated on batches of the new drug substance manufactured by the proposed commercial process, allowing sufficient latitude to deal with normal manufacturing and analytical variation and the stability characteristics of the new drug substance. Although normal manufacturing variations are expected, significant variation in batch-to-batch impurity levels can indicate that the manufacturing process of the new drug substance is not adequately controlled and validated. The use of two decimal places for thresholds (See Table 2-1) does not necessarily indicate the precision of the acceptance criteria for specified impurities and total impurities.

In summary, the new drug substance specification should include, where applicable, the following list of impurities:

Organic Impurities

- Each specified identified impurity
- Each specified unidentified impurity
- Any unspecified impurity with an acceptance criterion of not more than (≤) the identification threshold
- Total impurities

Residual Solvents

Inorganic Impurities

5. Qualification of impurities

Qualification is the process of acquiring and evaluating data that establishes the biological safety of an individual impurity or a given impurity profile at the level(s) specified. The applicant should provide a rationale for establishing impurity acceptance criteria that includes safety considerations. The level of any impurity present in a new drug substance that has been adequately tested in safety and/or clinical studies would be considered qualified. Impurities that are also significant metabolites present in animal and/or human studies are generally considered qualified. A level of a qualified impurity higher than that present in a new drug substance can also be justified based on an analysis of the actual amount of impurity administered in previous relevant safety studies.

If data are unavailable to qualify the proposed acceptance criterion of an impurity, studies to obtain such data can be appropriate when the usual qualification thresholds given in Table 2-1 are exceeded.

Higher or lower thresholds for qualification of impurities can be appropriate for some individual drugs based on scientific rationale and level of concern, including drug class effects and clinical experience. For example, qualification can be especially important when there is evidence that such impurities in certain drugs or therapeutic classes have previously been associated with adverse reactions in patients. In these instances, a lower qualification threshold can be appropriate. Conversely, a higher qualification threshold can be appropriate for individual drugs when the level of concern for safety is less than usual based on similar considerations (e.g., patient population, drug class effects, clinical considerations). Proposals for alternative thresholds would be considered on a case-by-case basis.

The "Decision tree for identification and qualification" (Figure 2-1) describes considerations

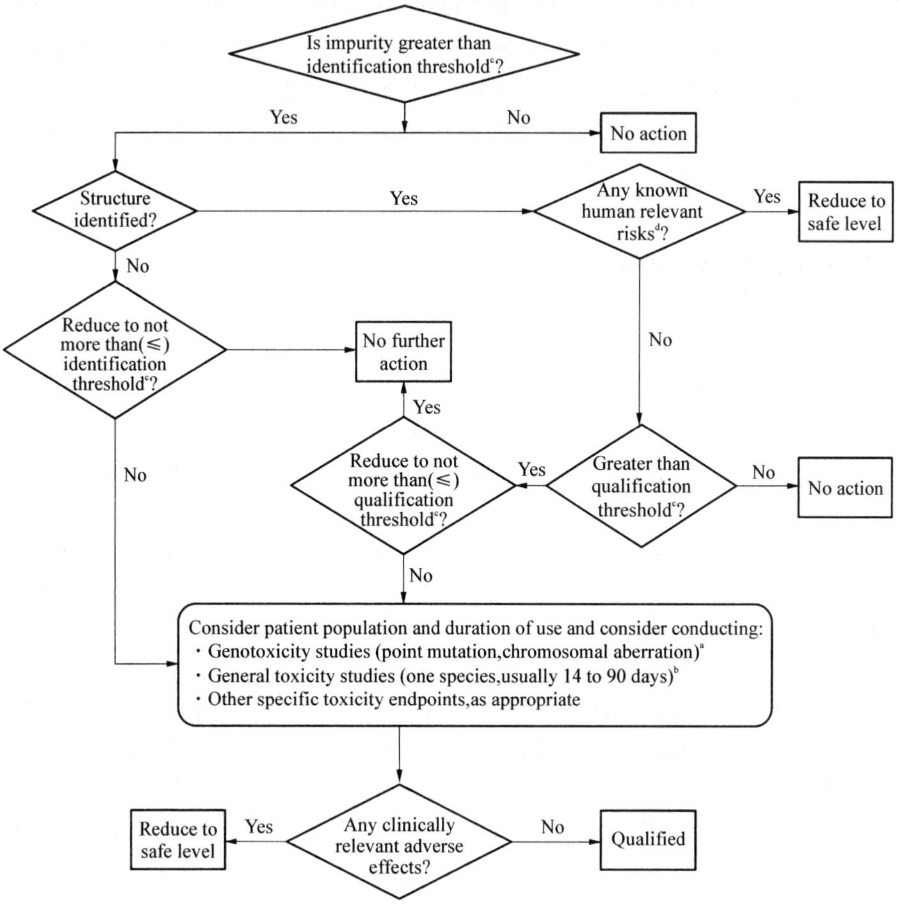

Figure 2-1 Decision tree for identification and qualification

for the qualification of impurities when thresholds are exceeded. In some cases, decreasing the level of impurity to not more than the threshold can be simpler than providing safety data. Alternatively, adequate data could be available in the scientific literature to qualify an impurity. If neither is the case, additional safety testing should be considered. The studies considered appropriate to qualify an impurity will depend on a number of factors, including the patient population, daily dose, and route and duration of drug administration. Such studies can be conducted on the new drug substance containing the impurities to be controlled, although studies using isolated impurities can sometimes be appropriate.

Notes on Figure 2－1: a. If considered desirable, a minimum screen (e.g., genotoxic potential), should be conducted. A study to detect point mutations and one to detect chromosomal aberrations, both *in vitro*, are considered an appropriate minimum screen. b. If general toxicity studies are desirable, one or more studies should be designed to allow comparison of unqualified to qualified material. The study duration should be based on available relevant information and performed in the species most likely to maximise the potential to detect the toxicity of an impurity. On a case-by-case basis, single-dose studies can be appropriate, especially for single-dose drugs. In general, a minimum duration of 14 days and a maximum duration of 90 days would be considered appropriate. c. Lower thresholds can be appropriate if the impurity is unusually toxic. d. For example, do known safety data for this impurity or its structural class preclude human exposure at the concentration present?

引自:ICH Expert Working Group. ICH harmonised tripartite guideline: impurities in new drug substances Q3A(R2)[R], 2006.

Words and Expressions

oligonucleotide	*n.* 寡(聚)核苷酸	radiopharmaceutical	*n.* 放射性药物
herbal	*adj.* 草药的	identification	*n.* 鉴定,确认
analytical	*adj.* 分析的	qualify	*v.* 鉴定
clinical	*adj.* 临床的,病房用的	volatile	*adj.* 挥发性的
by-product	*n.* 副产物	degradation	*n.* 降解
charcoal	*n.* 木炭	vehicle	*n.* 载体,传达媒介
suspension	*n.* 混悬剂	extraneous	*adj.* 外源性,外来的
contaminant	*n.* 污染物	polymorphic forms	多晶型
enantiomer	*n.* 对映异构体	appraisal	*n.* 评价,评估
raw material	起始原料,起始物料	batch	*n.* 批次,批量
criteria	*n.* 标准,规范,尺度	variation	*n.* 变更,变异
response factor	响应因子	threshold	*n.* 阈值
metabolite	*n.* 代谢产物	decimal	*n.* 小数;*adj.* 十进位的,小数的

precision	*n.* 精密，精确度，精确	pharmacopoeial	*adj.* 药典的
carry-over	(样品)遗留(如从上次实验中残留至下次实验)		

Notes

① new drug substance: 新原料药。先前尚未在任何成员国或地区注册的具有治疗作用的活性成分(也称为新分子或新化学实体)。它可以是某种已获批准的药物的一种复合物、简单的酯或盐。

② chemical development studies: 化学开发研究。对新原料药合成工艺进行放大、优化以及验证的研究。

③ enantiomeric impurity: 对映异构体杂质。与药物具有相同的分子式，但其分子中原子的空间排列不同并且为不能重合的镜像化合物。

④ extraneous contaminant: 外源污染物。来源于生产工艺以外的杂质。

⑤ herbal products: 草药。以植物和/或植物药制品为活性组分制成的药物制剂。在一些传统药中，可能也含无机物的材料或动物组织。

⑥ impurity profile: 杂质谱。对存在于某一新原料药中的已鉴定或未鉴定杂质的数量及含量进行分析。

⑦ polymorphic forms: 多晶型。某一药物的不同结晶形态。包括溶剂化或水合物(伪晶体及无定型)。

⑧ potential impurity: 潜在杂质。按照理论推测在生产或储存过程中可能产生的杂质。其在新原料药中可能存在，也可能不存在。

⑨ qualification: 界定。是获得和评价某些数据的过程，这些数据可用于确保单个杂质或在特定的含量下的一系列杂质的生物安全性。

⑩ starting material: 起始物料。在新原料药的合成中，作为一种成分结合到中间体或新原料药中的物质。起始物料通常市场上有供应，并具有确定的化学、物理性质和结构。

Exercises

1. Put the following into English:

草药	中间体	配体	试剂

2. Put the following into Chinese:

impurity profile	identified threshold	enantiomeric
polymorphic forms	qualificated threshold	starting material
specified impurity	extraneous contaminant	

Reading Materia:

History of Thalidomide

Thalidomide was first marketed to health professionals as a sedative. However, its use rapidly increased amongst pregnant women due to the drug's ability to alleviate morning sickness. Soon after its rise in popularity, medical professionals began to note a series of congenital mutations in children born from mothers who used the drug during pregnancy, resulting in the thalidomide tragedy known today. In more contemporary times, thalidomide has become renowned for its effectiveness in treating malignant and inflammatory diseases.

What is thalidomide?

Thalidomide was first developed by CIBA, a Swiss pharmaceutical company in the early 1950s, and subsequently introduced as Contergan by Chemi Grunenthal. The drug was initially advertised as a sedative which would allow users to undergo a deep sleep in the absence of a hangover and with a reduced risk of developing drug dependency. At the time, basic testing was done on the drug, and it was considered not to have any toxic effects on humans. However, unlike today's level of rigorous testing, the drug was not analyzed for any potentially dangerous teratogenic effects. Following its release, the drug became popular as a morning sickness remedy for pregnant women due to its anti-emetic effects. This increase in use for pregnant women was aided by the fact that the drug could be obtained without a prescription, and was affordable. However, following its widespread use in Japan, Australia, and Europe, practitioners began to notice links between mothers who had taken thalidomide and the presence of congenital mutations in their children.

Thalidomide and congenital mutations

In the 1960s, two medical professionals: Dr Widukind Lenze and Dr William McBride, observed an association between the use of thalidomide in expecting mothers and congenital malformations. Once publicized, these findings were further backed by several cases across the globe with a reported 10, 000 children thought to have been born with phocomelia. As a consequence, thalidomide was removed from the market in the majority of countries in 1961, with some still offering the drugs several years after. Across the cases of patients, a range of abnormalities was noted including:

a. Phocomelia—a congenital deformity whereby the hands and feet are bound to the child's trunk, absent or grossly underdeveloped.

b. Disfigurements of the ear.

c. Ocular abnormalities.

d. Facial palsies.

e. Internal organ damage.

f. Congenital heart disease.

In addition to these abnormalities, there was an increase in the number of miscarriages reported by women during the same period of time.

Due to the fact the US Food and Drug Administration didn't approve or license the use of thalidomide, no congenital disabilities were reported in America. Frances Kelsey was credited for avoiding the tragedy as she was worried about the implications the drug would have on pregnant women and the overall safety of the drug due to incidences of peripheral neuropathy. In response to the thalidomide tragedy, the pharmaceutical industry has taken great strides to increase the rigor of testing before introducing any new drugs onto the market. Specifically, a set of requirements were introduced for all drug testing to investigate potential developmental toxicity.

Contemporary uses of thalidomide

Since the tragedy, there's been an increased body of research investigating the action of thalidomide. It was found that thalidomide acts through two main types of mechanisms: anti-angiogenesis and anti-inflammatory. These properties have made the drug an ideal treatment option for a range of medical conditions.

Thalidomide as a treatment for multiple myeloma

Back in 2006, thalidomide became one of the first new agents in over ten years to be approved to treat plasma cell myeloma, a type of bone marrow cancer. Due to boasting anti-angiogenic properties, thalidomide prevents the metastasis, growth, and hypervascularity of cancer tumors.

Thalidomide as a treatment for leprosy

Soon after its withdrawal, thalidomide was prescribed to treat a specific implication of leprosy: erythema nodosum leprosum (ENL). ENL is an immunological implication thought to occur in approximately half of those with lepromatous leprosy. It is characterized by the presence of a range of conditions including neuritis, orchitis, lymphadenitis, and eye inflammation. As ENL is an inflammatory condition, thalidomide's anti-inflammatory properties make it a viable treatment option. A lot has been learned, and changes have been made since the thalidomide tragedy. The drug is no longer recommended for use in pregnant women.

However, due to more in-depth research and increased knowledge surrounding the drug's mechanisms of action, it is now being used to treat and manage a range of health conditions safely.

引自: BENNETT, C. History of Thalidomide[EB/OL]. (2023-03-15)[2023-07-18]. https://www.news-medical.net/health/History-of-Thalidomide.aspx.

Unit 2

Guideline on the Need for Carcinogenicity Studies of Pharmaceuticals

1. Introduction

The objectives of carcinogenicity studies are to identify a tumorigenic potential in animals and to assess the relevant risk in humans. Any cause for concern derived from laboratory investigations, animal toxicology studies, and data in humans may lead to a need for carcinogenicity studies. The practice of requiring carcinogenicity studies in rodents was instituted for pharmaceuticals that were expected to be administered regularly over a substantial part of a patient's lifetime. The design and interpretation of the results from these studies preceded much of the available current technology to test for genotoxic potential and the more recent advances in technologies to assess systemic exposure. These studies also preceded our current understanding of tumorigenesis with non-genotoxic agents. Results from genotoxicity studies, toxicokinetics, and mechanistic studies can now be routinely applied in preclinical safety assessment. These additional data are important not only in considering whether to perform carcinogenicity studies but for interpreting study outcomes with respect to relevance for human safety. Since carcinogenicity studies are time consuming and resource intensive, they should only be performed when human exposure warrants the need for information from life-time studies in animals in order to assess carcinogenic potential.

2. Historical background

In Japan, according to the 1990 "Guidelines for Toxicity Studies of Drugs Manual", carcinogenicity studies were needed if the clinical use was expected to be continuously for 6 months or longer. If there was cause for concern, pharmaceuticals generally used continuously for less than 6 months may have needed carcinogenicity studies. In the United States, most pharmaceuticals were tested in animals for their carcinogenic potential before widespread use in humans. According to the US Food and Drug Administration, pharmaceuticals generally used for 3 months or more required carcinogenicity studies. In Europe, the Rules Governing Medicinal Products in the European Community defined the circumstances when carcinogenicity studies were required. These circumstances included administration over a substantial period of life, i.e., continuously during a minimum period of 6 months or frequently in an intermittent manner so that the total exposure was similar.

3. Objective of the guideline

The objective of this guideline is to define the conditions under which carcinogenicity studies, should be conducted to avoid the unnecessary use of animals in testing, and to provide consistency in worldwide regulatory assessments of applications. It is expected that these studies will be performed in a manner that reflects currently accepted scientific standards.

The fundamental considerations in assessing the need for carcinogenicity studies are the maximum duration of patient treatment and any perceived cause for concern arising from other investigations. Other factors may also be considered such as the intended patient population, prior assessment of carcinogenic potential, the extent of systemic exposure, the (dis)similarity to endogenous substances, the appropriate study design, or the timing of study performance relative to clinical development.

4. Factors to consider for carcinogenicity testing

4.1 Duration and exposure

Carcinogenicity studies should be performed for any pharmaceutical whose expected clinical use is continuous for at least 6 months (Note: It is expected that most pharmaceuticals indicated for 3 months treatment would also likely be used for 6 months. In an inquiry to a number of pharmaceutical research and regulatory groups, no cases were identified in which a pharmaceutical would be used only for 3 months.).

Certain classes of compounds may not be used continuously over a minimum of 6 months but may be expected to be used repeatedly in an intermittent manner. It is difficult to determine and to justify scientifically what time represents a clinically relevant treatment periods for frequent use with regard to carcinogenic potential, especially for discontinuous treatment periods. For pharmaceuticals used frequently in an intermittent manner in the treatment of chronic or recurrent conditions, carcinogenicity studies are generally needed. Examples of such conditions include allergic rhinitis, depression, and anxiety. Carcinogenicity studies may also need to be considered for certain delivery systems which may result in prolonged exposures. Pharmaceuticals administered infrequently or for short duration of exposure (e.g., anaesthetics and radiolabelled imaging agents) do not need carcinogenicity studies unless there is cause for concern.

4.2 Cause for concern

Carcinogenicity studies may be recommended for some pharmaceuticals if there is concern about their carcinogenic potential. Criteria for defining these cases should be very carefully considered because this is the most important reason to conduct carcinogenicity studies for most categories of pharmaceuticals. Several factors which could be considered may include: (1) previous demonstration of carcinogenic potential in the product class that is considered relevant to humans; (2) structure-activity relationship suggesting carcinogenic risk; (3) evidence of preneoplastic lesions in repeated dose toxicity studies; and (4) long-term tissue retention of parent compound or metabolite(s) resulting in local tissue reactions or other pathophysiological responses.

4.3 Genotoxicity

Unequivocally genotoxic compounds, in the absence of other data, are presumed to be trans-species carcinogens, implying a hazard to humans. Such compounds need not be subjected to long-term carcinogenicity studies. However, if such a drug is intended to be administered chronically to humans a chronic toxicity study (up to one year) may be necessary to detect early tumorigenic effects.

Assessment of the genotoxic potential of a compound should take into account the totality of the findings and acknowledge the intrinsic value and limitations of both *in vitro* and *in vivo* tests. The test battery approach of *in vitro* and *in vivo* tests is designed to reduce the risk of false negative results for compounds with genotoxic potential. A single positive result in any assay for genotoxicity does not necessarily, mean that the test compound poses a genotoxic hazard to humans.

4.4 Indication and patient population

When carcinogenicity studies are required, they usually need to be completed before application for marketing approval. However, completed rodent carcinogenicity studies are not needed in advance of the conduct of large scale clinical trials, unless there is special concern for the patient population.

For pharmaceuticals developed to treat certain serious diseases, carcinogenicity testing need not be conducted before market approval although these studies should be conducted post-approval. This speeds the availability of pharmaceuticals for life-threatening or severely debilitating diseases, especially where no satisfactory alternative therapy exists.

In instances where the life-expectancy in the indicated population is short (i.e., less than 2–3 years) no long-term carcinogenicity studies may be required. For example, oncolytic agents intended for treatment of advanced systemic disease do not generally need carcinogenicity studies. In cases where the therapeutic agent for cancer is generally successful and life is significantly prolonged, there may be later concerns regarding secondary cancers. When such pharmaceuticals are intended for adjuvant therapy in tumour free patients or for prolonged use in noncancer indications, carcinogenicity studies are usually needed.

4.5 Route of exposure

The route of exposure in animals should be the same as the intended clinical route when feasible. If similar metabolism and systemic exposure can be demonstrated by differing routes of administration, then carcinogenicity studies should only be conducted by a single route, recognizing that it is important that relevant organs for the clinical route (e.g., lung for inhalational agents) be adequately exposed to the test material. Evidence of adequate exposure may be derived from pharmacokinetic data.

4.6 Extent of systemic exposure

Pharmaceuticals applied topically (e.g., dermal and ocular routes of administration) may need carcinogenicity studies. Pharmaceuticals showing poor systemic exposure from topical routes in humans may not need studies by the oral route to assess the carcinogenic potential to internal organs. Where there is cause for concern for photocarcinogenic potential carcinogenicity studies by dermal application (generally in mice) may be needed. Pharmaceuticals administered by the ocular route may not require carcinogenicity studies unless there is cause for concern or unless there is significant systemic exposure.

For different salts, acids, or bases of the same therapeutic moiety, where prior carcinogenicity studies are available, evidence should be provided that there are no significant changes in pharmacokinetics, pharmacodynamics, or toxicity. When changes in exposure and consequent toxicity are noted, then additional bridging studies may be used to determine whether additional carcinogenicity studies are needed. For esters and complex derivatives, similar data would be valuable in assessing the need for an additional carcinogenicity study, but this should be considered on a case-by-case basis.

4.7 Endogenous peptides and protein substances or their analogs

Endogenous peptides or proteins and their analogs, produced by chemical synthesis, by extraction/purification from an animal/human source or by biotechnological methods such as recombinant DNA technology may require special considerations.

Carcinogenicity studies are not generally needed for endogenous substances given essentially as replacement therapy (i.e., physiological levels), particularly where there is previous clinical experience with similar products (for example, animal insulins, pituitary-derived growth hormone, and calcitonin).

Although not usually necessary, long-term carcinogenicity studies in rodent species should be considered for the other biotechnology products noted above, if indicated by the treatment duration, clinical indication, or patient population (providing neutralizing antibodies are not elicited to such an extent in repeated dose studies as to invalidate the results). Conduct of carcinogenicity studies may be important in the following circumstances: (1) for products where there are significant differences in biological effects to the natural counterpart(s); (2) for products where modifications lead to significant changes in structure compared to the natural counterpart; and (3) for products resulting in humans in a significant increase over the existing local or systemic concentration (i.e., pharmacological levels).

5. Need for additional testing

The relevance of the results obtained from animal carcinogenicity studies for assessment of human safety are often cause for debate. Further research may be needed, investigating the mode of action, which may result in confirming the presence or the lack of carcinogenic potential for humans. Mechanistic studies are useful to evaluate the relevance of tumour findings in animals for human safety.

引自:ICH Expert Working Group. ICH harmonised tripartite guideline: guideline on the need for carcinogenicity studies of pharmaceuticals S1A[R], 1995.

Words and Expressions

carcinogenicity	*n.* 致癌性,致癌作用	tumorigenic	*adj.* 致瘤的,致癌性
rodent	*n.* 啮齿动物	genotoxic	*n.* 基因毒性,遗传毒性
toxicokinetics	*n.* 毒物代谢动力学	intermittent	*adj.* 间歇性的
endogenous	*adj.* 内源性的	duration	*n.* 持续时间,用药期限
exposure	*n.* 暴露量	chronic	*adj.* 慢性的, 习惯性的
recurrent	*adj.* 复发的,周期性的	allergic rhinitis	变应(过敏)性鼻炎
depression	*n.* 忧郁	anxiety	*n.* 焦虑
delivery systems	给(递)药系统	anaesthetics	*n.* 麻醉剂
radiolabelled imaging agents	放射性同位素标记造影剂	parent compound	原型化合物
local tissue	局部组织	pathophysiological	*adj.* 病理生理
indication	*n.* 适应证	alternative therapy	替代疗法
life-expectancy	预期寿命	adjuvant therapy	辅助疗法
inhalational	*adj.* 吸入的	pharmacokinetic	*n.* 药物代谢动力学
dermal	*adj.* 皮肤的	ocular	*n.* 眼睛;*adj.* 眼睛的,视觉的
oral	*n.* 口试;*adj.* 口部的	photocarcinogenic	*adj.* 光致癌的
calcitonin	*n.* 降钙素	pituitary	*n.* 脑垂体
pharmacodynamics	*n.* 药效动力学,药效学		

Notes

① carcinogenicity studies: 致癌性试验。致癌试验是检验外来化合物及其代谢物是否具有诱发癌或肿瘤作用。致癌试验检验的对象包括恶性肿瘤(癌)和良性肿瘤。致癌试验一般可分为长期致癌试验和短期快速筛检法,其中长期致癌试验多于哺乳动物中进

行,一般多用大鼠、小鼠等啮齿动物,如条件许可,尚可于狗和猴等一种非啮齿动物中进行。用完整哺乳动物进行的长期致癌试验结果可靠,但试验过程较长,不能在较短时间内得出结论,费用也较高,故近年来有许多短期快速方法正在发展,较常用方法有致突变试验法、哺乳动物细胞体外转化试验、DNA 修复合成试验。

② genotoxicity studies: 遗传毒性试验。遗传毒性试验又称为致突变试验,可分为体外试验和整体动物实验,目的是检测药物是否会通过不同机理直接或间接引起基因突变,为药物的安全性评价提供依据。遗传毒性试验常用的方法是采用不同检测机理的遗传毒性试验的组合,即体内试验和体外试验相结合、原核细胞与真核细胞相结合、不同试验终点相结合,从而提高了检测具有遗传毒性药物的能力。

③ toxicokinetics: 毒物代谢动力学。简称毒代动力学,是药物代谢动力学原理在毒理学中的应用。在毒性试验条件下,研究大于治疗剂量的药物在毒理实验动物体内的吸收、分布、代谢和排泄的过程及其随时间的动态变化规律,阐明药物或其代谢产物在体内的部位、数量和毒性作用间的关系。

④ the Rules Governing Medicinal Products in the European Community: 欧盟药品管理规则。

⑤ Guidelines for Toxicity Studies of Drugs Manual: 药物毒性研究指导原则手册。

⑥ preneoplastic lesions: 癌前病变。癌前病变是指继续发展下去具有癌变可能的某些病变,如黏膜白斑、交界痣、慢性萎缩性胃炎、结直肠的多发性腺瘤性息肉、某些良性肿瘤等。

Exercises

1. Put the following into English:

致癌性　　暴露量　　用药期限　　遗传毒性　　替代疗法

2. Put the following into Chinese:

carcinogenicity studies	endogenous substance	discontinuous treatment
structure-activity relationship	local tissue reaction	chronic toxicity study
false negative	route of exposure	

Unit 3
Process Validation: General Principles

1. Introduction

The lifecycle concept links product and process development, qualification of the commercial manufacturing process, and maintenance of the process in a state of control during routine commercial production. This guidance supports process improvement and innovation through sound science.

This guidance covers the following categories of drugs:

- Human drugs
- Veterinary drugs
- Biological and biotechnology products
- Finished products and active pharmaceutical ingredients (APIs or drug substances)
- The drug constituent of a combination (drug and medical device) product

This guidance does not cover the following types of products:

- Type A medicated articles and medicated feed
- Medical devices
- Dietary supplements
- Human tissues intended for transplantation regulated under section 361 of the Public Health Service Act

2. Background

In the *Federal Register* of May 11, 1987 (52 FR 17638), FDA issued a notice announcing the availability of a guidance entitled *Guideline on General Principles of Process Validation* (the 1987 guidance). Since then, we have obtained additional experience through our regulatory oversight that allows us to update our recommendations to industry on this topic. This revised guidance conveys FDA's current thinking on process validation and is consistent with basic principles first introduced in the 1987 guidance. The revised guidance also provides recommendations that reflect some of the goals of FDA's initiative entitled "Pharmaceutical CGMPs for the 21st Century—A Risk-Based Approach," particularly with regard to the use of technological advances in pharmaceutical manufacturing, as well as implementation of modern risk management and quality system tools and concepts. This revised guidance replaces the 1987 guidance.

FDA has the authority and responsibility to inspect and evaluate process validation performed by manufacturers. The CGMP regulations for validating pharmaceutical (drug) manufacturing require that drug products be produced with a high degree of assurance of meeting all the attributes they are intended to possess [21 CFR 211.100(a) and 211.110(a)].

2.1 Process validation and drug quality

Effective process validation contributes significantly to assuring drug quality. The basic principle of quality assurance is that a drug should be produced that is fit for its intended use. This principle incorporates the understanding that the following conditions exist:

- Quality, safety, and efficacy are designed or built into the product.
- Quality cannot be adequately assured merely by in-process and finished-product inspection or testing.
- Each step of a manufacturing process is controlled to assure that the finished product meets all quality attributes including specifications.

2.2 Approach to process validation

For purposes of this guidance, **process validation** is defined as the collection and evaluation of data, from the process design stage through commercial production, which establishes scientific evidence that a process is capable of consistently delivering quality product. Process validation involves a series of activities taking place over the lifecycle of the product and process. This guidance describes process validation activities in three stages.

- Stage 1—Process Design: The commercial manufacturing process is defined during this stage based on knowledge gained through development and scale-up activities.
- Stage 2—Process Qualification: During this stage, the process design is evaluated to determine if the process is capable of reproducible commercial manufacturing.
- Stage 3—Continued Process Verification: Ongoing assurance is gained during routine production that the process remains in a state of control.

This guidance describes activities typical of each stage, but in practice, some activities might occur in multiple stages.

Before any batch from the process is commercially distributed for use by consumers, a manufacturer should have gained a high degree of assurance in the performance of the manufacturing process such that it will consistently produce APIs and drug products meeting those attributes relating to identity, strength, quality, purity, and potency. The assurance should be obtained from objective information and data from laboratory-, pilot-, and/or commercial-scale studies. Information and data should demonstrate that the commercial manufacturing process is capable of consistently producing acceptable quality products within commercial manufacturing conditions.

A successful validation program depends upon information and knowledge from product and process development. This knowledge and understanding is the basis for establishing an approach to control of the manufacturing process that results in products with the desired quality attributes. Manufacturers should:

- Understand the sources of variation
- Detect the presence and degree of variation
- Understand the impact of variation on the process and ultimately on product attributes
- Control the variation in a manner commensurate with the risk it represents to the process and product

Each manufacturer should judge whether it has gained sufficient understanding to provide a high degree of assurance in its manufacturing process to justify commercial distribution of the product. Focusing exclusively on qualification efforts without also understanding the manufacturing process and associated variations may not lead to adequate assurance of quality. After establishing and confirming the process, manufacturers must maintain the process in a state of control over the life of the process, even as materials, equipment, production environment, personnel, and manufacturing procedures change.

Manufacturers should use ongoing programs to collect and analyze product and process data to evaluate the state of control of the process. These programs may identify process or product problems or opportunities for process improvements that can be evaluated and implemented through some of the activities described in Stages 1 and 2.

Manufacturers of legacy products can take advantage of the knowledge gained from the original process development and qualification work as well as manufacturing experience to continually improve their processes. Implementation of the recommendations in this guidance for legacy products and processes would likely begin with the activities described in Stage 3.

3. Statutory and regulatory requirements for process validation

Process validation for drugs (finished pharmaceuticals and components) is a legally enforceable requirement under section 501(a)(2)(B) of the Act [21 U.S.C. 351(a)(2)(B)], which states the following:

A drug...shall be deemed to be adulterated...if...the methods used in, or the facilities or controls used for, its manufacture, processing, packing, or holding do not conform to or are not operated or administered in conformity with current good manufacturing practice to assure that such drug meets the requirements of this Act as to safety and has the identity and strength, and meets the quality and purity characteristics, which it purports or is represented to possess.

FDA regulations describing current good manufacturing practice (CGMP) for finished pharmaceuticals are provided in 21 CFR parts 210 and 211.

The CGMP regulations require that manufacturing processes be designed and controlled to assure that in-process materials and the finished product meet predetermined quality requirements and do so consistently and reliably. Process validation is required, in both general and specific terms, by the CGMP regulations in parts 210 and 211. The foundation for process validation is provided in § 211.100(a), which states that "[t]here shall be written procedures for production and process control **designed to assure** that the drug products have the identity, strength, quality, and purity they purport or are represented to possess..." (emphasis added). This regulation requires manufacturers to design a process, including operations and controls, which results in a product meeting these attributes.

Other CGMP regulations define the various aspects of validation. For example, § 211.110(a), "Sampling and testing of in-process materials and drug products", requires that control procedures "...be established to monitor the output and to validate the performance of those manufacturing processes that may be responsible for causing variability in the characteristics of in-process material and the drug product" (emphasis added). Under this regulation, even well-designed processes must include in-process control procedures to assure final product quality. In addition, the CGMP regulations regarding sampling set forth a number of requirements for validation: samples must represent the batch under analysis [§ 211.160(b)(3)]; the sampling plan must result in statistical confidence [§ 211.165(c) and (d)]; and the batch must meet its predetermined specifications [§ 211.165(a)].

In addition to sampling requirements, the CGMP regulations also provide norms for establishing in-process specifications as an aspect of process validation. Section 211.110(b) establishes two principles to follow when establishing in-process specifications. The first principle is that "...in-process specifications for such characteristics [of in-process material and the drug product] shall be consistent with drug product final specifications...." Accordingly, in-process material should be controlled to assure that the final drug product will meet its quality requirements. The second principle in this regulation further requires that in-process specifications "... shall be derived from previous acceptable process average and process variability estimates where possible and determined by the application of suitable statistical procedures where appropriate." This requirement, in part, establishes the need for manufacturers to analyze process performance and control batch-to-batch variability.

The CGMP regulations also describe and define activities connected with process design, development, and maintenance. Section 211.180(e) requires that information and data about product quality and manufacturing experience be periodically reviewed to determine whether any changes to the established process are warranted. Ongoing feedback about product quality and process performance is an essential feature of process maintenance.

In addition, the CGMP regulations require that facilities in which drugs are manufactured be of suitable size, construction, and location to facilitate proper operations (§ 211.42). Equipment must be of appropriate design, adequate size, and suitably located to facilitate operations for its intended use (§ 211.63). Automated, mechanical, and electronic equipment must be calibrated, inspected, or checked according to a written program designed to assure proper performance (§ 211.68).

In summary, the CGMP regulations require that manufacturing processes be designed and controlled to assure that in-process materials and the finished product meet predetermined quality requirements and do so consistently and reliably.

引自：U.S. Department of Health and Human Services, Food and Drug Administration, Center for Drug Evaluation and Research, et al. Process validation: general principles and practices[R], 2011.

Words and Expressions

process validation	工艺验证	veterinary	*n.* 兽医；*adj.* 医牲畜的，兽医的
finished product	制剂产品	medical device	医疗器械(设备)
dietary supplement	膳食补充剂	process design	工艺设计
process qualification	工艺评价	adulterate	*v.* 掺杂，掺假
facility	*n.* 设备	holding	*n.* 贮藏
sampling plan	取样方案	confidence	*n.* 置信区间

Notes

① Commercial manufacturing process: 商品化生产工艺。本指南中，商品化生产工艺这一专业名词指生产出商品化产品的生产工艺(即用于经销、流通、出售或拟出售的药品)，不包括临床试验或用于治疗的研究型药物(IND)材料。

② *Guideline on General Principles of Process Validation* (the 1987 guidance)：工艺验证的一般指导原则(1987 年版指南)。1987 年的指南由一个以医疗器械与放射卫生中心(CDRH)为代表的工作组起草撰写。从那以后，CDRH 选择与全球医疗器械法规协调组织(Global Harmonization Task Force, GHIF)合作发布了自己的工艺验证指南，该文件即：质量管理体系-工艺验证第 2 版，其中的原则与建议对于药品生产工艺验证也很有帮助。

③ "Pharmaceutical CGMPs for the 21st Century—A Risk-Based Approach,"："21 世纪制药业现行药品生产规范——一种基于风险的方法，"。此处"，"表示该规范有多个报告，如"Pharmaceutical CGMPs for the 21st Century—A Risk-Based Approach, Second Progress Report and Implementation Plan"，"Pharmaceutical CGMPs for the 21st Century—A Risk-Based Approach, Final Report"等。

④ CGMPs: 动态药品生产管理规范。Current Good Manufacture Practices 的缩写，也译作现行药品生产管理规范。CGMP 是美、欧、日等国执行的 GMP 规范，是国际药品生产管理标准。该规范强调现场管理(current)，要求在产品生产和物流的全过程都必须验证，主要目的是保证稳定的产品质量，药品质量就是 CGMP 的核心，而实现这一目标的过程(或理解为现场)是最重要的。

Exercises

Put the following into Chinese:

a. Process validation is defined as the collection and evaluation of data, from the process design stage through commercial production, which establishes scientific evidence that a process is capable of consistently delivering quality product.

b. A drug...shall be deemed to be adulterated...if...the methods used in, or the facilities or controls used for, its manufacture, processing, packing, or holding do not conform to or are not operated or administered in conformity with current good manufacturing practice to assure that such drug meets the requirements of this Act as to safety and has the identity and strength, and meets the quality and purity characteristics, which it purports or is represented to possess.

Reading material:

The Cutter Incident—1955

Poliomyelitis is a peculiarly nasty, highly contagious disease of the central nervous system that all too often leaves its victims permanently disabled with withered limbs. Polio epidemics swept through postwar America with alarming frequency, growing more virulent every year. An epidemic of 1952 killed over 3, 000 people and left more than 21, 000 with varying degrees of paralysis. So when medical researcher Jonas Salk developed a viable vaccine, he was hailed as a savior.

There followed the largest medical experiment in history, involving 1, 800, 000 children in a rigorous double-blind trial. On April 12, 1955 the vaccine was hailed a success: it was deemed safe and effective. The public health authorities immediately licensed several pharmaceutical companies to produce the vaccine in bulk so they could carry out a mass immunization program to rid America of this latter-day plague. Among the chosen companies was Cutter Laboratories of Berkeley, California. A terrible mishap in their laboratory led to what is known as the Cutter incident.

Cutter manufactured 120, 000 doses of vaccine in which the polio virus had not been deactivated properly. In two out of eight batches produced at the laboratory, some of the polio virus had survived the formaldehyde treatment designed to kill it. It was one of the worst disasters in the history of the US pharmaceutical industry. Cutter was sued and, when the case came to court in 1958, although the firm was cleared of negligence, it was ordered to pay damages for breach of warranty—having claimed that the vaccine was safe when it manifestly wasn't. The company never again produced polio vaccine and in the 1970s was taken over by German pharmaceutical giant Bayer, the third largest pharmaceutical company in the world.

引自：DEVASTATING DISASTERS. The Cutter Incident-1955[EB/OL].(2015-06)[2022-12-24]. https://devastatingdisasters.com/health/34902139/2015/06/the-cutter-incident-1955/.

Unit 4

The Recommended Stages of Process Qualification

During the process qualification (PQ) stage of process validation, the process design is evaluated to determine if it is capable of reproducible commercial manufacture. This stage has two elements: (1) design of the facility and qualification of the equipment and utilities and (2) process performance qualification (PPQ). During Stage 2, CGMP-compliant procedures must be followed. Successful completion of Stage 2 is necessary before commercial distribution. Products manufactured during this stage, if acceptable, can be released for distribution.

1. Design of a facility and qualification of utilities and equipment

Proper design of a manufacturing facility is required under part 211, subpart C, of the CGMP regulations on *Buildings and Facilities*. It is essential that activities performed to assure proper facility design and commissioning precede PPQ. Here, the term **qualification** refers to activities undertaken to demonstrate that utilities and equipment are suitable for their intended use and perform properly. These activities necessarily precede manufacturing products at the commercial scale.

Qualification of utilities and equipment generally includes the following activities:

- Selecting utilities and equipment construction materials, operating principles, and performance characteristics based on whether they are appropriate for their specific uses.
- Verifying that utility systems and equipment are built and installed in compliance with the design specifications (e. g., built as designed with proper materials, capacity, and functions, and properly connected and calibrated).
- Verifying that utility systems and equipment operate in accordance with the process requirements in all anticipated operating ranges. This should include challenging the equipment or system functions while under load comparable to that expected during routine production. It should also include the performance of interventions, stoppage, and start-up as is expected during routine production. Operating ranges should be shown capable of being held as long as would be necessary during routine production.

Qualification of utilities and equipment can be covered under individual plans or as part of an overall project plan. The plan should consider the requirements of use and can incorporate risk management to prioritize certain activities and to identify a level of effort in both the performance and documentation of qualification activities. The plan should identify the following items:

a. The studies or tests to use,

b. The criteria appropriate to assess outcomes,

c. The timing of qualification activities,

d. The responsibilities of relevant departments and the quality unit, and

e. The procedures for documenting and approving the qualification.

The project plan should also include the firm's requirements for the evaluation of changes. Qualification activities should be documented and summarized in a report with conclusions that address criteria in the plan. The quality control unit must review and approve the qualification plan and report (§ 211.22).

2. Process performance qualification

The process performance qualification (PPQ) is the second element of Stage 2, process qualification. The PPQ combines the actual facility, utilities, equipment (each now qualified), and the trained personnel with the commercial manufacturing process, control procedures, and components to produce commercial batches. A successful PPQ will confirm the process design and demonstrate that the commercial manufacturing process performs as expected.

Success at this stage signals an important milestone in the product lifecycle. A manufacturer must successfully complete PPQ before commencing commercial distribution of the drug product. The decision to begin commercial distribution should be supported by data from commercial-scale batches. Data from laboratory and pilot studies can provide additional assurance that the commercial manufacturing process performs as expected.

The approach to PPQ should be based on sound science and the manufacturer's overall level of product and process understanding and demonstrable control. The cumulative data from all relevant studies (e.g., designed experiments; laboratory, pilot, and commercial batches) should be used to establish the manufacturing conditions in the PPQ. To understand the commercial process sufficiently, the manufacturer will need to consider the effects of scale. However, it is not typically necessary to explore the entire operating range at commercial scale if assurance can be provided by process design data. Previous credible experience with sufficiently similar products and processes can also be helpful. In addition, we strongly recommend firms employ objective measures (e.g., statistical metrics) wherever feasible and meaningful to achieve adequate assurance.

In most cases, PPQ will have a higher level of sampling, additional testing, and greater scrutiny of process performance than would be typical of routine commercial production. The level of monitoring and testing should be sufficient to confirm uniform product quality throughout the batch. The increased level of scrutiny, testing, and sampling should continue through the process verification stage as appropriate, to establish levels and frequency of routine sampling and monitoring for the particular product and process. Considerations for the duration of the heightened sampling and monitoring period could include, but are not limited to, volume of production, process complexity, level of process understanding, and experience with similar products and processes.

The extent to which some materials, such as column resins or molecular filtration media, can be re-used without adversely affecting product quality can be assessed in relevant laboratory studies. The usable lifetimes of such materials should be confirmed by an ongoing PPQ protocol during commercial manufacture.

A manufacturing process that uses PAT may warrant a different PPQ approach. PAT processes are designed to measure in real time the attributes of an in-process material and then adjust the process in a timely control loop so the process maintains the desired quality of the output material. The process design stage and the process qualification stage should focus on the measurement system and control loop for the measured attribute. Regardless, the goal of validating any manufacturing process is the same: to establish scientific evidence that the process is reproducible and will consistently deliver quality products.

3. PPQ protocol

A written protocol that specifies the manufacturing conditions, controls, testing, and expected outcomes is essential for this stage of process validation. We recommend that the protocol discuss the following elements:

- The manufacturing conditions, including operating parameters, processing limits, and component (raw material) inputs.
- The data to be collected and when and how it will be evaluated.
- Tests to be performed (in-process, release, characterization) and acceptance criteria for each significant processing step.
- The sampling plan, including sampling points, number of samples, and the frequency of sampling for each unit operation and attribute. The number of samples should be adequate to provide sufficient statistical confidence of quality both within a batch and between batches. The confidence level selected can be based on risk analysis as it relates to the particular attribute under examination. Sampling during this stage should be more extensive than is typical during routine production.
- Criteria and process performance indicators that allow for a science- and risk-based decision about the ability of the process to consistently produce quality products. The criteria should include:

—A description of the statistical methods to be used in analyzing all collected data (e.g., statistical metrics defining both intra-batch and inter-batch variability).

—Provision for addressing deviations from expected conditions and handling of nonconforming data. Data should not be excluded from further consideration in terms of PPQ without a documented, science-based justification.

- Design of facilities and the qualification of utilities and equipment, personnel training and qualification, and verification of material sources (components and container/closures), if not previously accomplished.

- Status of the validation of analytical methods used in measuring the process, in-process materials, and the product.
- Review and approval of the protocol by appropriate departments and the quality unit.

4. PPQ protocol execution and report

Execution of the PPQ protocol should not begin until the protocol has been reviewed and approved by all appropriate departments, including the quality unit. Any departures from the protocol must be made according to established procedure or provisions in the protocol. Such departures must be justified and approved by all appropriate departments and the quality unit before implementation (§ 211.100).

The commercial manufacturing process and routine procedures must be followed during PPQ protocol execution [§ 211.100(b) and § 211.110(a)]. The PPQ lots should be manufactured under normal conditions by the personnel routinely expected to perform each step of each unit operation in the process. Normal operating conditions should include the utility systems (e.g., air handling and water purification), material, personnel, environment, and manufacturing procedures.

A report documenting and assessing adherence to the written PPQ protocol should be prepared in a timely manner after the completion of the protocol. This report should:

- Discuss and cross-reference all aspects of the protocol.
- Summarize data collected and analyze the data, as specified by the protocol.
- Evaluate any unexpected observations and additional data not specified in the protocol.
- Summarize and discuss all manufacturing nonconformances such as deviations, aberrant test results, or other information that has bearing on the validity of the process.
- Describe in sufficient detail any corrective actions or changes that should be made to existing procedures and controls.
- State a clear conclusion as to whether the data indicates the process met the conditions established in the protocol and whether the process is considered to be in a state of control. If not, the report should state what should be accomplished before such a conclusion can be reached. This conclusion should be based on a documented justification for the approval of the process, and release of lots produced by it to the market in consideration of the entire compilation of knowledge and information gained from the design stage through the process qualification stage.
- Include all appropriate department and quality unit review and approvals.

引自：U.S. Department of Health and Human Services, Food and Drug Administration, Center for Drug Evaluation and Research, et al. Process validation: general principles and practices[R], 2011.

Words and Expressions

design of the facility	(厂房)设施设计	utility	*n.* 公共设施
process performance qualification	工艺性能确认	verifying	*n.* 核实
design specification	设计规范	capacity	*n.* 产能
calibrate	*v.* 校准	stoppage	*n.* 停止，中止
operating range	运行范围	pilot study	中试研究
statistical metrics	统计指标	volume of production	产量
column resins	树脂柱	molecular filtration media	分子过滤介质
operating parameter	运行参数	processing limit	工艺限度
air handling	空气处理	deviation	*n.* 偏差

Notes

PAT: Process Analytical Technologies,过程分析技术。过程分析技术是指以保证终产品质量为目的,通过对有关原料、生产中物料及工艺的关键参数和性能指标进行实时(即在工艺过程中)检测的一个集设计、分析和生产控制为一体的系统。

Exercises

Put the following into Chinese:

a. This stage has two elements: (1) design of the facility and qualification of the equipment and utilities and (2) process performance qualification (PPQ). During Stage 2, CGMP-compliant procedures must be followed.

b. Here, the term qualification refers to activities undertaken to demonstrate that utilities and equipment are suitable for their intended use and perform properly. These activities necessarily precede manufacturing products at the commercial scale.

c. The PPQ combines the actual facility, utilities, equipment (each now qualified), and the trained personnel with the commercial manufacturing process, control procedures, and components to produce commercial batches.

d. The sampling plan, including sampling points, number of samples, and the frequency of sampling for each unit operation and attribute. The number of samples should be adequate to provide sufficient statistical confidence of quality both within a batch and between batches. The confidence level selected can be based on risk analysis as it relates to the particular attribute under examination. Sampling during this stage should be more extensive than is typical during routine production.

Unit 5
Rare Pediatric Disease Priority Review Vouchers, Questions and Answers

1. Background and overview

Section 529 of the FD&C Act is intended to encourage development of new drug and biological products ("drugs") for prevention and treatment of certain rare pediatric diseases. Although there are existing incentive programs to encourage the development and study of drugs for rare diseases, pediatric populations, and unmet medical needs, section 529 provides an additional incentive for rare pediatric diseases, which may be used alone or in combination with other incentive programs. These other incentive programs include: orphan-drug designation and the associated benefits under the Orphan Drug Act for rare disease therapies; programs that encourage or require the study of drugs used in pediatric populations under the Best Pharmaceuticals for Children Act (BPCA) and the Pediatric Research Equity Act (PREA); and various programs to facilitate and expedite development and review of new drugs to address unmet medical needs in the treatment of serious or life-threatening conditions. Even so, Congress has recognized that there remain unmet medical needs among patients with rare diseases that occur primarily in pediatric populations. By enacting section 529, Congress intended to stimulate new drug development for rare pediatric diseases by offering additional incentives for obtaining FDA approval of these products.

Under section 529, the sponsor of a human drug application [as defined in section 735(1) of the FD&C Act] for a rare pediatric disease drug product may be eligible for a voucher that can be used to obtain a priority review for a subsequent human drug application submitted under section 505(b)(1) of the FD&C Act or section 351 of the Public Health Service (PHS) Act after the date of approval of the rare pediatric disease drug product.

2. Definitions, policies, and procedures—questions and answers

Q1. What is a "rare pediatric disease"?

"**Rare pediatric disease**" is defined at section 529(a)(3) as a disease that:

- "primarily affects individuals aged from birth to 18 years, including age groups often called neonates, infants, children, and adolescents," which we (throughout this document, the terms "we" and "FDA" interchangeably) interpret as meaning that greater than 50% of the affected population in the U.S. is aged 0 through 18 years; **and**

- Is "a rare disease or condition" as defined in section 526, which includes diseases and conditions that affect fewer than 200, 000 persons in the United States (U.S.) and diseases and conditions that affect a larger number of persons and for which there is no reasonable expectation that the costs of developing and making available the drug in the U.S. can be recovered from sales of the drug in the U.S.

Of note, section 529 describes the pediatric population as from birth through 18 years. This age range differs from how FDA defines the pediatric population in other contexts. Generally, for drug and biological products, FDA considers the pediatric population to include patients from birth through 16 years.

Under our interpretation of section 529(a)(3), a drug would qualify as a drug for a "rare pediatric disease" if the entire prevalence of the disease or condition in the U.S. is below 200, 000 and if more than 50% of patients with the disease are 0 through 18 years of age. Another way a drug may qualify as a drug for a "rare pediatric disease" is if it is for an "orphan subset" of a disease or condition that otherwise affects 200, 000 or more persons in the U.S., and if this subset is primarily (i.e., more than 50%) comprised of individuals aged 0 through 18 years.

The calculation of prevalence estimates will depend on whether the drug is a therapeutic drug or a vaccine, diagnostic drug, or preventive drug, as follows:

- **For therapeutic drugs,** prevalence estimates of the entire affected U.S. population and of those aged 0 through 18 years should be based on the number of individuals diagnosed with the disease or condition. For some diseases and conditions, individuals may have an underlying genetic abnormality at birth but may not develop manifestations of the disease until later, if ever. In these instances, whether individuals are considered "diagnosed" for the purpose of estimating prevalence may depend on whether the product is intended to treat an underlying genetic abnormality, attenuate or prevent progression of the clinical expression of the disease, or treat the clinical symptoms or manifestations of the disease.
- **For vaccines, diagnostic drugs, and preventive drugs**, prevalence estimates should be based on the number of persons of all ages, and those 0 through 18 years of age, to whom the drug will be administered in the U.S. annually.

Qualifying as a drug for a "rare pediatric disease" is not sufficient to receive a priority review voucher. For sponsors to receive such a voucher, the application for the drug must meet all of the remaining eligibility criteria described in response to Question 2.

Q2. What is a "rare pediatric disease product application"?

The term "**rare pediatric disease product application**" is defined in section 529(a)(4) of the FD&C Act. It refers to an application that:

- Is a human drug application as defined in section 735(1) of the FD&C Act:

—For prevention or treatment of a "rare pediatric disease";

—That contains no active ingredient (including any ester or salt of the active ingredient) that has been previously approved in any other application under section 505(b)(1), 505(b)(2), or 505(j) of the FD&C Act or section 351(a) or 351(k) of the PHS Act.

- That FDA deems eligible for priority review.
- Is submitted under section 505(b)(1) of the FD&C Act or section 351(a) of the Public Health Service Act.
- Relies on clinical data derived from studies examining a pediatric population and dosages of the drug intended for that population.
- Does not seek approval for an adult indication in the original rare pediatric disease product application; and
- Is approved after the date of enactment of the Prescription Drug User Fee Amendments of 2012 (July 9, 2012).

Q3. What is the rare pediatric disease designation process?

Under section 529(d), a sponsor may choose to request rare pediatric disease designation. This designation process is entirely voluntary; requesting designation is not a prerequisite to requesting or receiving a priority review voucher.

If sponsors choose to request such designation, section 529(d)(2) provides that they shall do so "at the same time" that they submit a request for orphan-drug designation under section 526 or a request for fast track designation under section 506.

Note that, while a request for rare pediatric disease designation may be submitted at the same time as a request for orphan-drug designation or fast track designation, each request should be submitted as a separate proposal (i.e., they should not be submitted in one combined package).

We remind sponsors of the timing for orphan-drug and fast track designation requests:

Timing of Requests for Orphan-Drug Designation: Under section 526, orphan-drug designation requests must be submitted before the sponsor's filing of a marketing application for the drug for the orphan use.

Timing of Requests for Fast Track Designation: Requests for fast track designation may be submitted at the time of original submission of the investigational new drug (IND) application or any time thereafter prior to receiving marketing approval of the NDA or BLA, although FDA encourages that such requests be submitted no later than the sponsor's pre-NDA/BLA meeting because many of the benefits of fast track designation will no longer be applicable after that time.

If sponsors submit a timely request for rare pediatric disease designation, section 529(d)(3) directs FDA to make a decision on the request no later than 60 days after submission. The statute directs FDA to decide whether to designate the drug as a drug for a "rare pediatric disease" and whether to designate the application for the drug as "a rare pediatric disease product application, " as described in response to Question "How will FDA respond to such designation requests?".

FDA recognizes that some sponsors may wish to submit a rare pediatric disease designation request at a different time—for example, if they had already submitted requests for orphan-drug and/or fast track designation before the enactment of FDASIA, or if for whatever reason they have no interest in submitting either such request but do want to submit a rare pediatric disease designation request. FDA is willing to accept designation requests submitted at a different time than that provided by statute as long as FDA receives the designation request before FDA has filed the NDA/BLA for the drug for the relevant indication. Although we will aim to respond to such requests in a timely manner, the 60-day response deadline does not apply. We will not accept requests for rare pediatric disease designation received after FDA has already filed the NDA/BLA for the drug for the relevant indication.

Even if sponsors have requested and received rare pediatric disease designation, they should include a request for a rare pediatric disease priority review voucher in their original NDA/BLA submission (either in the initial package sent or up until the point of NDA/BLA filing). Sponsors who have not requested designation, or have not received designation, may also request a voucher in their original NDA/BLA submission.

Q4. What is a priority review?

When a marketing application receives priority review designation, FDA's goal is to take action on the application within 6 months as compared to 10 months for a standard review. An application for a drug is eligible for a priority review, absent the use of a priority review voucher, if it is determined that the drug treats a serious condition and, if approved, would provide a significant improvement in safety or effectiveness. Also eligible for priority review are: any supplement that proposes a labeling change pursuant to a report on a pediatric study under section 505A; an application for a drug that has been designated as a qualified infectious disease product; or an application or supplement for a drug submitted with a priority review voucher.

Q5. Will a drug that receives rare pediatric disease designation also qualify for orphan-drug designation?

We anticipate that many rare pediatric disease drug products will qualify for designation as orphan drugs (if such designation is sought) because a "rare pediatric disease" also must be a "rare disease or condition" as defined in section 526, including those that affect fewer than 200, 000 persons in the U.S. There are instances, however, where a drug may qualify as a drug for a "rare pediatric disease" but not qualify for orphan-drug designation, or vice versa, as explained below. The following examples illustrate situations in which a drug might receive rare pediatric disease designation but not also immediately qualify for orphan drug designation.

- Assume that a drug receives "rare pediatric disease" designation but is considered the "same drug" under the orphan drug regulations as an already approved drug for the same orphan use. 21 CFR 316.3(b)(14). This drug would not be eligible to receive orphan-drug designation absent a plausible hypothesis that it may be clinically superior to the already approved drug. 21 CFR 316.20(a) and (b)(5). **Note**: Even though this drug may receive "rare pediatric disease" designation, the application for the drug likely would not qualify as an "application for a rare pediatric disease product application"—and hence not be likely to receive a priority review voucher—because, if the drug is considered the "same drug" as a previously approved drug under the orphan drug regulations, then it will generally contain a previously approved active ingredient (including any ester or salt of the active ingredient), rendering it ineligible for a voucher under section 529.

- Assume a sponsor plans to develop a drug for a rare pediatric disease but so far has very little data suggesting that the drug may be effective in that disease (e.g., only *in vitro* data supporting the drug's mechanism of action in a related disease). It is possible that this level of data may suffice for rare pediatric disease designation but generally it would not suffice for orphan-drug designation. This is because, to qualify for orphan-drug designation, an applicant must supply sufficient information to establish a medically plausible basis for expecting the drug to be effective in the prevention, diagnosis, or treatment of the rare disease or condition. The sponsor may eventually obtain orphan designation for the drug after developing or obtaining more supportive data for use of the drug for the rare disease or condition, including *in vivo* and/or clinical data in the rare disease or condition.

If a drug receives orphan-drug designation, it may be eligible for orphan-drug exclusivity, tax credits for qualified clinical testing, orphan product grant funding, as well as fee exemptions under section 736 of the FD&C Act.

引自:U.S. Department of Health and Human Services, Food and Drug Administration, Center for Biologics Evaluation and Research, et al. Rare pediatric disease priority review vouchers[R], 2014.

Words and Expressions

pediatric	*adj.* 小儿科的	designation	*n.* 认定
priority review	优先评审	voucher	*n.* 凭证
orphan-drug	*n.* 孤儿药	neonate	*n.* 新生儿
diagnose	*v.* 诊断	clinical symptom	临床症状
verifying	*n.* 核实	prevalence	*n.* 发(患)病率
adolescent	*n.* 青少年;*adj.* 青春期的,青少年的	infectious	*adj.* 有传染性的,易传染的
rare pediatric disease	罕见儿科疾病		

Notes

① FD&C Act: Federal Food, Drug, and Cosmetic Act,《联邦食品、药品和化妆品法案》。

② Orphan Drug Act:《孤儿药法案》,简称 ODA。美国于 1983 年颁布《孤儿药法案》,逐步建立鼓励罕见病药物研发的长效机制。

③ Best Pharmaceuticals for Children Act:《最佳儿童药品法案》。2002 年,美国国会通过了针对儿科用药研究制定的专门性法律规范——《最佳儿童药品法案》。该法案延续了 1997 年《食品药品监督管理局现代化法案》中设立的 6 个月的儿科独占保护期政策,成立儿科治疗办公室负责伦理和上市后的安全性问题。

④ Pediatric Research Equity Act:《儿科研究平等法案》。美国国会于 2003 年颁布该法案规定企业除非经 FDA 同意准予豁免,在 NDA 或 BLA 递交之前必须提供儿科临床研究启动计划(Initial Pediatric Study Plan)所需要的安全性和有效性评价资料。

⑤ Human Drug Application (as defined in section 735(1) of the FD&C Act): "人类药物申请"这个术语在 FD&C 法第 735(1)节的法定定义是(A)根据本篇第 355(b)节提交的新药审批申请,或(B)根据第 42 篇第 262 节(a)小节提交的生物制品许可证申请。这个术语并不包括这类申请的补充申请;不包括用于输注的全血或成分血的申请;不包括涉及 1992 年 9 月 1 日之前许可的局部应用小牛血制品、过敏原性提取物产品,或根据第 42 篇第 26 节许可的体外诊断生物制品之申请;不包括涉及 1992 年 9 月 1 日之前审批的大容量胃肠外药品的申请;不包括单纯为继续生产而提交的生物制品许可证申请;不包括州或联邦政府实体提交的不用于商业销售的药物之申请或补充申请。但这个术语包括(B)小段描述的,用于静脉应用或输注的单剂注射大容量生物制品的许可证申请。

⑥ Public Health Service (PHS) Act:《公共卫生服务法》。又称"美国检疫法",于 1994 年通过。该法有关防范传染病方面的规定内容主要包括:明确严重传染病的界定程序,制定传染病控制条例,规定检疫官员的职责,同时对来自特定地区的人员和货物,及有关检疫站、检疫场所与港口管理,民航与民航飞机的检疫等均做出了详尽的规定。此外还对战争时期的特殊检疫进行了规范。

⑦ Orphan Subset: 罕见临床亚型。"罕见临床亚型"需要证明因该药的一个或多个特性,如药物毒性、作用机制或该药的既往临床经验,在该关注亚型之外(患有该疾病或状况的其余个人)使用该药并不合适。一种药物可根据两种可能的"罕见临床亚型"证据而有资格成为用于"罕见儿科疾病"的药物:(A)一种非罕见疾病或状况的"罕见临床亚型",前提是该亚型发病率低于 200, 000 的界限,而且发病者超过 50%年龄为 0~18 岁;或(B)若该药物治疗的罕见疾病或状况不是主要累及 0~18 岁的个体,但由于该药的一个或多个特征,该药物主要用于某个年龄范围,则是受累年龄范围的"罕见临床亚型"。

⑧ IND: Investigational New Drug,新药临床试验申请。

⑨ NDA: New Drug Application,新药生产上市注册申请。

⑩ BLA: Biologics License Application,生物制品许可申请。

⑪ FDASIA: Food and Drug Administration Safety and Innovation Act,《食品药品监督管理局安全与创新法案》。该法案于 2012 年 7 月 9 日正式生效,涉及处方药、医疗器械、仿制药、生物仿制药生产企业的用户费用。该法案授权 FDA 向企业收取用户费用以支持创新型新药、医疗器械仿制药和相似生物制品(生物仿制药)的评审,该法案也重新授权了两个项目,鼓励儿科药物的研发。

Exercises

1. Put the following into English:

孤儿药　　优先评审　　快速通道　　传染病　　罕见儿科疾病

2. Put the following into Chinese:

a. "primarily affects individuals aged from birth to 18 years, including age groups often called neonates, infants, children, and adolescents, " which we (throughout this document, the terms "we" and "FDA" interchangeably) interpret as meaning that greater than 50% of the affected population in the U.S. is aged 0 through 18 years.

b. Assume that a drug receives "rare pediatric disease" designation but is considered the "same drug" under the orphan drug regulations as an already approved drug for the same orphan use. This drug would not be eligible to receive orphan-drug designation absent a plausible hypothesis that it may be clinically superior to the already approved drug.

Reading material:

Brittle Bones, Unbreakable Spirit

Sam Renke has been challenging misconceptions about disability since she was diagnosed with a rare genetic condition called Osteogenesis Imperfecta (OI), or brittle bone condition, shortly after her birth.

"I was born with multiple fractures and I actually fractured in utero, " explains the award-winning actress, broadcaster, writer, disability rights campaigner and diversity and inclusion consultant. After Sam was taken to another hospital for further tests, her parents received the diagnosis.

OI is caused by a genetic mutation that affects the production of collagen, causing bones to break easily, as well as bone deformities and other symptoms depending on the severity of the condition. The UK's Brittle Bone Society estimates that approximately 1 in 15, 000 people are born with OI. Diagnosis is made with the aid of X-rays and genetic tests, among other medical technologies.

"From the word go, it was very much a negative narrative, like 'Sorry, there's something wrong with your child.' I think it was quite traumatic for my parents in the sense of being young themselves, having an older child with no disability, and then this kind of bombshell," the former teacher tells This Is MedTech. At first Sam went to a school for disabled children, but it soon became apparent that she would be better served by a mainstream education.

Nevertheless, she faced barriers that many disabled people experience. "I've always been labelled as someone that wouldn't achieve or couldn't achieve. I think a lot of my success comes from the drive to prove people wrong," she notes. But it wasn't until years later when she became a trustee for the Brittle Bone Society that Sam really started having pride in her disability.

"I don't see my wheelchair or my disability as a negative. I see it as a positive. I'm not wheelchair bound. I'm a wheelchair user," Sam points out. She wants to change societal attitudes towards disability and works with numerous charities to raise awareness and challenge inequalities.

Having been in and out of hospital throughout her life, Sam acknowledges the important role medtech plays in supporting disabled people. "I had telescopic rods implanted in my legs at two and four years old. Later I had spinal fusion. These, as well as hydrotherapy and physiotherapy, were integral to my health and prolonging my life," she says.

As for everyday support, Sam believes the conversation around technology for disabled people is especially important. To this end, she welcomes the greater focus on aesthetics and design that she's seeing these days. "As a young disabled woman, I want to feel sexy and confident. I want my wheelchairs and any kind of mobility aid to reflect my character and not say 'medical device'."

Sam also sees the potential for technology to foster more independence. "The pandemic has shown that we can all work and we can all be more inclusive. We've proven that with the way we can use virtual meetings and so forth. I think we need to be working with disabled people and using technology to that advantage and maybe not looking at things in such a linear way."

引自:FINN K. Brittle bones, unbreakable spirit[EB/OL].(2021-11-03)[2022-07-20].https://www.thisismedtech.com/brittle-bones-unbreakable-spirit/.

Chapter 3
Drugs Production

Unit 1
Production of Drugs

Depending on their production or origin, pharmaceutical agents can be split into three groups:

Ⅰ. Totally synthetic materials (synthetics),

Ⅱ. Natural products, and

Ⅲ. Products from partial syntheses (semi-synthetic products).

The emphasis of the present book is on the most important compounds of groups Ⅰ and Ⅲ - thus Drug synthesis. This does not mean, however, that natural products or other agents are less important. They can serve as valuable lead structures, and they are frequently needed as starting materials or as intermediates for important synthetic products.

Table 3-1 gives an overview of the different methods for obtaining pharmaceutical agents.

Table 3-1 Possibilities for the preparation of drugs

Methods	Examples
1. Total synthesis	—over 75% of all pharmaceutical agents (synthetics)
2. Isolation from natural sources (natural products):	
2.1 Plants	—alkaloids; enzymes; heart glycosides; polysaccharides; tocopherol; steroid precursors (diosgenin, sitosterin); citral (intermediate product for vitamins A, E, and K)
2.2 Animal organs	—enzymes; peptide hormones; cholic acid from gall; insulin from the pancreas; sera and vaccines
2.3 Other sources	—cholesterol from wool oils; L-amino acids from keratin and gelatine hydrolysates
3. Fermentation	—antibiotics; L-amino acids; dextran; targeted modifications on steroids, e. g, 11-hydroxylation; also insulin, interferon, antibodies, peptide hormones, enzymes, vaccines
4. Partial synthetic modification of natural products (semisynthetic agents):	—alkaloid compounds; semisynthetic β-lactam antibiotics; steroids; human insulin

Several therapeutically significant natural products which were originally obtained from natural sources are today more effectively-i.e. more economically-prepared by total synthesis. Such examples include L-amino acids, Chloramphenicol, Caffeine, Dopamine, Epinephrine,

Levodopa, peptide hormones, Prostaglandins, D-Penicillamine, Vincamine, and practically all vitamins.

Over the last few years, fermentation–i.e. microbiological processes–has become extremely important. Through modern technology and results from genetic selection leading to the creation of high performance mutants of microorganisms, fermentation has already become the method of choice for a wide range of substances. Both Eukaryonts (yeasts and moulds) and Prokaryonts (single bacterial cells, and actinomycetes) are used as microorganisms. The following product types can be obtained:

1. cell material (single cell protein),
2. enzymes,
3. primary degradation products (primary metabolites),
4. secondary degradation products (secondary metabolites).

Disregarding the production of dextran from the mucous membranes of certain microorganisms, e.g. *Leuconostoc mesenteroides,* classes 2 and 3 are the relevant ones for the preparation of drugs. Dextran itself, with a molecular weight of 50, 000–100, 000, is used as a blood plasma substitute. Among the primary metabolites, the L-amino acids from mutants of *Corynebacterium glutamicum* and *Brevibacterium flavum* are especially interesting. From these organisms, some 350, 000 tones of monosodium L-glutamate (food additive) and some 70, 000 tones of L-lysine (supplement for vegetable proteins) are produced. Further important primary metabolites are the purina nucleotides, organic acids, lactic acid, citric acid, and vitamins, for example vitamin B_{12} from *Propionibacteriw shermanii.*

Among the secondary metabolites the antibiotics must be mentioned first. The following five groups represent a yearly worldwide value of US $ 17 billion:

Penicillins (*Penicillium chrysogenum*),
Cephalosporins (*Cephalosporium acremonium*),
Tetracyclines (*Streptoryces aureofaciens*),
Erythromycins (*Streptomyces erythreus*),
Aminoglycosides (e.g. streptomycin from *Streptomyces griseus*).

About 5, 000 antibiotics have already been isolated from microorganisms, but of these only somewhat fewer than 100 are in therapeutic use. It must be remembered, however, that many derivatives have been modified by partial synthesis for therapeutic use; some 50, 000 agents have been semisynthetically obtained from β-lactams alone in the last decade. Fermentations are carried out in stainless steel fermentors with volumes up to 400 m^3. To avoid contamination of the microorganisms with phages etc, the whole process has to be performed under sterile conditions. Since the more important fermentations occur exclusively under aerobic conditions, a good supply of oxygen or air (sterile) is needed. Carbon dioxide sources include carbohydrates, e.g. molasses, saccharides, and glucose. Additionally, the microorganisms must be supplied in the growth medium with nitrogen-containing compounds such as ammonium sulfate, ammonia, or urea, as well as with inorganic phosphates. Furthermore, constant optimal pH and temperature are required. In the case of penicillin G, the fermentation is finished after 200 hours, and the cell mass is separated by filtration. The desired active agents are isolated from the filtrate by absorption or extraction processes. The cell mass, if not the desired product, can be further used as an animal feedstuff owing to its high protein content.

By modern recombinant techniques, microorganisms have been obtained which also allow production of peptides which were not encoded in the original genes. Modified E. *coli* bacteria make it thus possible to produce A- and B-chains of human insulin or proinsulin analogs. The disulfide bridges are formed selectively after isolation, and the final purification is effected by chromatographic procedures. In this way, human insulin is obtained totally independently from any pancreatic material taken from animals.

Other important peptides, hormones, and enzymes, such as human growth hormone (HGH), neuroactive peptides, somatostatin, interferons, tissue plasminogen activator(TPA), lymphokines, calcium regulators like calmodulin, protein vaccines, as well as monoclonal antibodies used as diagnostics, are synthesized in this way.

The enzymes or enzymatic systems which are present in a single microorganism can be used for directed stereospecific and regiospecific chemical reactions. This principle is especially useful in steroid chemistry. Here we may refer only to the microbiological 11-α-hydroxylation of progesterone to 11-α-hydroxyprogesterone, a key product used in the synthesis of cortisone. Isolated enzymes are important today not only because of the technical importance of the enzymatic saccharification of starch, and the isomerization of glucose to fructose. They are also significant in the countless test procedures used in diagnosing illness, and in enzymatic analysis which is used in the monitoring of therapy.

A number of enzymes are themselves used as active ingredients. Thus preparations containing proteases (e.g. chymotrypsin, pepsin, and trypsin), amylases and lipases, mostly in combination with synthetic antacids, promote digestion. Streptokinase and urokinase are important in thrombolytics, and asparaginase is used as a cytostatic agent in the treatment of leukemia.

Finally, mention must be made of the important use of enzymes as 'biocatalysts' in chemical reactions where their stereospecificity and selectivity can be used. Known examples are the enzymatic cleavage of racemates of N-acetyl-D, L-amino acids to give L-amino acids, the production of 6-aminopenicillanic acid from benzylpenicillin by means of penicillinamidase and the aspartase-catalysed stereospecific addition of ammonia to fumaric acid in order to produce L-aspartic acid.

In these applications, the enzymes can be used in immobilized forms—somehow bound to carriers—and so used as heterogeneous catalysts. This is advantageous because they can then easily be separated from the reaction medium and recycled for further use.

Another important process depending on the specific action of proteases is applied for the production of semisynthetic human insulin. This starts with pig insulin in which the alanine in the 30-position of the B-chain is replaced by a threonine *tert*-butyl ester by the selective action of trypsin. The insulin ester is separated, hydrolyzed to human insulin and finally purified by chromatographic procedures.

Sources for enzymes include not only microorganisms but also vegetable and animal materials.

In Table 3−1, it was already shown that over 75% of all pharmaceutical agents are obtained by total synthesis. Therefore, knowledge of the synthetic routes is useful. Understanding also makes it possible to recognize contamination of the agents by intermediates and by-products. For the reason of effective quality control, the registration authorities in many countries demand as essentials for registration a thorough documentation on the production process. Knowledge of drug syntheses provides the R&D chemist with valuable stimulation as well.

There are neither preferred structural classes for all pharmaceutically active compounds nor preferred reaction types. This implies that practically the whole field of organic and in part also organometallic chemistry is covered. Nevertheless, a larger number of starting materials and intermediates are more frequently used, and so it is useful to know the possibilities for their preparation from primary chemicals. For this reason, it is appropriate somewhere in this book to illustrate a tree of especially important intermediates. These latter intermediates are the key compounds used in synthetic processes leading to an enormous number of agents. For the most part chemicals are involved which are produced in large amounts. In a similar way this is also true for the intermediates based on the industrial aromatic compounds toluene, phenol and chlorobenzene. Further key compounds may be shown in a table which can be useful in tracing cross-relationships in syntheses.

In addition to the actual starting materials and intermediates, solvents are required both as a reaction medium and for purification via recrystallization. Frequently used solvents are methanol, ethanol, isopropanol, butanol, acetone, ethyl acetate, benzene, toluene and xylene. To a lesser extent diethyl ether, tetrahydrofuran (THF), glycol ethers, dimethylformamide (DMF) and dimethyl sulphoxide (DMSO) are used in special reactions.

Reagents used in larger amounts are not only acids (hydrochloric acid, sulfuric acid, nitric acid, acetic acid) but also inorganic and organic bases (sodium hydroxide, potassium hydroxide, potassium carbonate, sodium bicarbonate, ammonia, tri-ethylamine, pyridine). Further auxiliary chemicals include active charcoal and catalysts. All of these supplementary chemicals (like the intermediates) can be a source of impurities in the final product.

In 1969, the WHO published a treatise on "Safeguarding Quality in Drugs". Appendix 2 is concerned with the "Proper Practice for Reparation and Safeguarding Quality in Drugs" (WHO Technical Report No. 418, 1969, Appendix 2; No. 567, 1975, Appendix lA). This has in the meantime become known as "Good Manufacturing Practices" or GMP rules, and these should now be obeyed in drug production. They form the basis for mutual recognition of quality certificates relating to the production of pharmaceuticals and for inspections of the production facilities.

For a long time, the US drug authority, the Food and Drug Administration (FDA), has issued regulations for the preparation of drugs analogous to the WHO rules, and it applies these strictly. Exports of drugs to the USA, like those of finished products, require regular inspection of the production facilities by the FDA.

It may merely be noted here that such careful control applies not only to the products, but also to the raw materials (control of starting materials), and also to the intermediates. Clearly the technical and hygienic equipment of the production and the storage areas have to fulfill set conditions.

Since only a few compounds, such as acetylsalicylic acid, paracetamol and vitamins, are prepared in large amounts, most of the actual production takes place in multi-purpose (multi-product) facilities. Special care has to be taken to avoid cross-contamination by other products what can be effected by good cleansing of used apparatus. A careful description and definition of all stored intermediates and products is needed.

Words and Expressions

alkaloid	*n.* 生物碱	polysaccharide	*n.* 多糖,多聚糖
precursor	*n.* 前体	steroid	*n.* 甾体
gall	*n.* 胆汁	hormone	*n.* 激素,荷尔蒙
insulin	*n.* 胰岛素	pancreas	*n.* 胰腺
serum	*n.* 血浆	vaccine	*n.* 疫苗,牛痘疫苗
cholesterol	*n.* 胆固醇	gelatin/gelatin	*n.* 骨胶,明胶
interferon	*n.* 干扰素	antibiotic	*adj.* 抗生的,抗菌的; *n.* 抗生素,抗菌素
fermentation	*n.* 发酵	antibody	*n.* 抗体
therapeutical	*adj.* 治疗(学)的	caffeine	*n.* 咖啡因,咖啡碱
yeast	*n.* 酵母	dopamine	*n.* 多巴胺(一种神经递质)
Penicillin	*n.* 青霉素	mucous	*adj.* 黏液的,分泌黏液的;*n.* 黏膜
Streptomycin	*n.* 链霉素	plasma	*n.* 血浆,淋巴液,等离子体
derivative	*n.* 衍生物;*adj.* 衍生的	sterile	*adj.* 不能生育的,无细菌的
feedstuff	*n.* 饲料	aerobic	*adj.* 需氧的,有氧的

starch	*n.* 淀粉	lymph	*n.* 淋巴,淋巴液
immobilize	*vt.* 固定化	streptokinase	*n.* 链球菌菌激酶
trypsin	*n.* 胰蛋白酶	glucose	*n.* 葡萄糖
intermediate	*n.* 中间体	heterogeneous	*adj.* 不均匀的,多相的
recrystallization	*n.* 重结晶	contamination	*n.* 玷污,污染,污染物
xylene	*n.* 二甲苯	hygienic	*adj.* 卫生学的,卫生的
toluene	*n.* 甲苯	extraction	*n.* 萃取,提取
ether	*n.* 醚	regiospecific reaction	区域专一性反应
benzene	*n.* 苯	stereospecific reaction	立体专一性反应

Notes

① Depending on their production or origin, pharmaceutical agents can be split into three groups.
本句子为被动句,根据被动语态的翻译技巧,结合中文表达方式与特点,可以将被动句翻译为主动句,在原主语前加译“把”。此外,从句中 their 指代 pharmaceutical agents。本句可翻译为:根据药剂的生产方法或者来源,可以把药剂分为三类。

② dextran: 右旋糖酐、葡聚糖。
葡聚糖是指以葡萄糖为单糖组成的同型多糖,葡萄糖单元之间以糖苷键连接。其中根据糖苷键的类型又可分为 alpha-葡聚糖和 beta-葡聚糖。alpha-葡聚糖中研究及使用较多的为 dextran,又称右旋糖酐,为一种多糖。存在于某些微生物在生长过程中分泌的黏液中。随着微生物种类和生长条件的不同,其结构也有差别。

③ Both Eukaryonts (yeasts and moulds) and Prokaryonts (single bacterial cells, and actinomycetes) are used as microorganisms. 其中 be used as 本意为“把……当作……用”“把……用作”,但是在本句中,不能如此翻译。因为 Eukaryonts 和 actinomycetes 自身是微生物,如果翻译为“把它们用作微生物”,很显然是错误的。由于英文写作特点,本文为了避免头重脚轻,把 as microorganisms 放在句子后面,句子原型应该是:As microorganisms, both Eukaryonts (yeasts and moulds) and Prokaryonts (single bacterial cells, and actinomycetes) are used.

④ FDA: 美国联邦食品及药物管理局(The Federal Food and Drng Administration)。

⑤ Additionally, the microorganisms must be supplied in the growth medium with nitrogen-containing compounds such as ammonium sulfate, ammonia, or urea, as well as with inorganic phosphates. 本句较长,成分较复杂,要正确翻译,必须理清句子成分。首先,本句谓语为动词词组 be supplied with,而 as well as 为连接并列成分的连词词组,其后面的 with 显然是与 supplied 连用,说明其为由 as well as 连接的省略句。该句骨架为:the microorganisms must be supplied with A as well as with B. 此外,in the growth medium

是修饰本句主语 the microorganisms，限定微生物为培养基中的微生物。

Exercises

1. Put the following into English:

生物碱	起始原料	重结晶	胆固醇
吡啶	甲苯	萃取	胰岛素醛

2. Put the following into Chinese:

polysaccharide	peptide	hormone	vaccine	heterogeneous catalyst
contamination	plasma	steroid	Penicillin	metabolite

Reading material A:

Sterile Products

Sterile products

Sterile products are dosage forms of therapeutic agents that are free of viable microorganisms. Principally, these include parenteral, ophthalmic, and irrigating preparations. Of these, parenteral products are unique among dosage forms of drugs because they are injected through the skin or mucous membranes into internal body compartment. Thus, because they have circumvented the highly efficient first line of body defense, the skin and mucous membranes, they must be free from microbial contamination and from toxic components as well as possess an exceptionally high level of purity. All components and processes involved in the preparation of these products must be selected and designed to eliminate, as much as possible, contamination of all types, whether of physical, chemical, or microbiologic origin.

Preparations for the eye, though not introduced into internal body cavities, are placed in contact with tissues that are very sensitive to contamination. Therefore, similar standards are required for ophthalmic preparations.

Irrigating solutions are now also required to meet the same standards as parenteral solutions because during an irrigation procedure, substantial amounts of these solutions can enter the bloodstream directly through open blood vessels of wounds or abraded mucous membranes. Therefore, the characteristics and standards presented in this chapter for the production of large-volume parenteral solutions apply equally to irrigating solutions.

Sterile products are most frequently solutions or suspensions, but may even be solid pellets for tissue implantation. The control of a process to minimize contamination for a small quantity of such a product can be achieved with relative ease. As the quantity of product increases, the problems of controlling the process to prevent contamination multiply. Therefore, the preparation of sterile products has become a highly specialized area in pharmaceutical processing. The standards established, the attitude of personnel, and the process control must be of a superior level.

Vehicles

By far the most frequently employed vehicle for sterile products is water, since it is the vehicle for all natural body fluids. The superior quality required for such use is described in the monograph on Water for Injection in the USP. Requirements may be even more stringent for some products, however.

One of the most inclusive tests for the quality of water is the total solids content, a gravimetric evaluation of the dissociated and undissociated organic and inorganic substances present in the water. However, a less time-consuming test, the electrolytic measurement of conductivity of the water, is the one most frequently used. Instantaneous measurements can be obtained by immersing electrodes in the water and measuring the specific conductance, a measurement that depends on the ionic content of the water. The conductance may be expressed by the meter scale as conductivity in micromhos, resistance in megohms, or ionic content as parts per million (ppm) of sodium chloride. The validity of this measurement as an indication of the purity of the water is inferential in that methods of producing high-purity water, such as distillation and reverse osmosis, can be expected to remove undissociated substances along with those that are dissociated. Undissociated substances such as pyrogens, however, could be present in the absence of ions and not be disclosed by the test. Therefore, for contaminants other than ions, additional tests should be performed.

Additional tests for quality of Water for Injection with permitted limits are described in the USP monographs. When comparing the total solids permitted for Water for Injection with that for Sterile Water for Injection, one will note that considerably higher values are permitted for Sterile Water for Injection. This is necessary because the latter product has been sterilized, usually by a thermal method, in a container that has dissolved to some extent in the water. Therefore, the solids content will be greater than for the nonsterilized product. On the other hand, the 10 ppm total solids officially permitted for Water for Injection may be much too high when used as the vehicle for many products. In practice, Water for Injection normally should not have a conductivity of more han 1 micromho (1 megohm, approximately 0.1 ppm NaCl).

Added substances

Substances added to a product to enhance its stability are essential for almost every product. Such substances include solubilizers, antioxidants, chelating agents, buffers, tonicity contributors, antibacterial agents, antifungal agents, hydrolysis inhibitors, antifoaming agents, and numerous other substances for specialized purposes. At the same time, these agents must be prevented from adversely affecting the product. In general, added substances must be nontoxic in the quantity administered to the patient. They should not interfere with the therapeutic efficacy nor with the assay of the active therapeutic compound. They must also be present and active when needed throughout the useful life of the product. Therefore, these agents must be selected with great care, and they must be evaluated as to their effect upon the entire formulation. An extensive review of excipients used in parenteral products and the means for adjusting pH of these products has recently been published and should be referred to for more detailed information.

Formulation

The formulation of a parenteral product involves the combination of one or more ingredients with a medicinal agent to enhance the convenience, acceptability, or effectiveness of the product. Rarely is it preferable to dispense a drug singly as a sterile dry powder unless the formulation of a stable liquid preparation is not possible.

On the other hand, a therapeutic agent is a chemical compound subject to the physical and chemical reactions characteristic of the class of compounds to which it belongs. Therefore, a careful evaluation must be made of every combination of two or more ingredients to ascertain whether or not adverse interactions occur, and if they do, of ways to modify the formulation so that the reactions are eliminated or minimized. The formulation of sterile products is challenging, therefore, to the knowledge and ingenuity of the persons responsible.

The amount of information available to the formulator concerning the physical and chemical properties of a therapeutic agent, particularly if it is a new compound, is often quite meager. Information concerning basic properties must be obtained, including molecular weight, solubility, purity, colligative properties, and chemical reactivity, before an intelligent approach to formulation can begin. Improvements in formulation are a continuing process, since important properties of a drug or of the total formulation may not become evident until the product has been stored or used for a prolonged time. However, because of the extensive test documentation required by the U.S. Food and Drug Administration (FDA), only outstanding formulations can be justified for continuance to the state of a marketed product.

Production

The production process includes all of the steps from the accumulation and combining of the ingredients of the formula to the enclosing of the product in the individual container for distribution. Intimately associated with these processes are the personnel who carry them out and the facilities in which they are performed. The most ideally planned processes can be rendered ineffective by personnel who do not have the right attitude or training, or by facilities that do not provide an efficiently controlled environment.

To enhance the assurance of successful manufacturing operation, all process steps must be carefully reduced to writing after being shown to be effective. These written process steps are often called standard operating procedures (SOPs). No extemporaneous changes are permitted to be made in these procedures; any change must go through the same approval steps as the original written SOP. Further, extensive records must be kept to give assurance at the end of the production process that all steps have been performed as prescribed, an aspect emphasized in the FDA's Good Manufacturing Practices. Such in-process control is essential to assuring the quality of the product, since these assurances are even more significant than those from product release testing. The production of a quality product is a result of the continuous, dedicated effort of the quality assurance, production, and quality control personnel within the plant in developing, performing, and confirming effective SOPs.

Words and Expressions

ophthalmic	*adj.* 眼的;*n.* 眼药	irrigate	*vt.* 冲洗
vehicles	*n.* 赋形剂,运载工具	gravimetric	*adj.* 重量分析的,重量的
cavity	*n.* 洞,腔	instantaneous	*adj.* 瞬间的,即刻的
abrade	*vt.* 磨,擦,擦伤	inferential	*adj.* 推论的,推理的
pellet	*n.* 药丸,小球	osmosis	*n.* 渗透,渗透作用
implantation	*n.* 对端植入法	dissociate	*vt.* 使分离,使离解
pyrogen	*n.* 致热质,热原	buffer	*vt.* 缓冲;*n.* 缓冲剂
solubilizing	*adj.* 增容的	tonicity	*n.* 强壮,强健
antioxidants	*n.* 抗氧剂	antifoaming	*adj.* 防沫的,消沫的
antifungal	*adj.* 抗真菌的	colligative	*adj.* 综合的,概括的
inhibitor	*n.* 抑制剂	accumulation	*n.* 积累,积聚物
wrap	*vt.* 包装	aseptic	*n.* 防腐剂;*adj.* 无菌的
specification	*n.* 说明书,规范,规格	extemporaneous	*adj.* 即席的,临时的

Notes

① preparation: 此处作制剂解释。

② ophthalmic preparation: 眼用制剂。

③ irrigating solution: 冲洗液。

④ megohm 为兆欧姆,micromhos 为微欧姆。

⑤ parts per million(ppm):百万分之一(10^{-6})。

⑥ Standard Operation Procedures(SOP):标准操作规程。

Exercises

1. Put the following into English:

灭菌产品	反向渗透	蒸馏	测量仪
电极	电导率	微生物	组织

2. Put the following into Chinese:

parenteral	ophthalmic	irrigating	microoganisms
contamination	specialize	conductivity	pycogens

Reading material B:

Death from Meningitis Outbreak Rises to 14 in US

Unit 2

Good Manufacturing Practice Guide for Active Pharmaceutical Ingredients

1. Introduction

1.1 Objective

This document (Guide) is intended to provide guidance regarding good manufacturing practice (GMP) for the manufacturing of active pharmaceutical ingredients (APIs) under an appropriate system for managing quality. It is also intended to help ensure that APIs meet the requirements for quality and purity that they purport or are represented to possess.

In this Guide "manufacturing" is defined to include all operations of receipt of materials, production, packaging, repackaging, labeling, relabeling, quality control (QC), release, storage and distribution of APIs, and the related controls. In this Guide the term "should" indicates recommendations that are expected to apply unless shown to be inapplicable or replaced by an alternative demonstrated to provide at least an equivalent level of quality assurance (QA). For the purposes of this Guide, the terms "current good manufacturing practices" and "good manufacturing practices" are equivalent.

The Guide as a whole does not cover safety aspects for the personnel engaged in the manufacture, nor aspects of protection of the environment. These controls are inherent responsibilities of the manufacturer and are governed by national laws.

This Guide is not intended to define registration/filing requirements or modify pharmacopoeial requirements. This Guide does not affect the ability of the responsible regulatory agency to establish specific registration/filing requirements regarding APIs within the context of marketing/manufacturing authorizations or drug applications. All commitments in registration/filing documents must be met.

1.2 Regulatory applicability

Within the world community, materials may vary as to the legal classification as an API. When a material is classified as an API in the region or country in which it is manufactured or used in a drug product, it should be manufactured according to this Guide.

1.3 Scope

This Guide applies to the manufacture of APIs for use in human drug (medicinal) products. It applies to the manufacture of sterile APIs only up to the point immediately prior to the APIs being rendered sterile. The sterilization and aseptic processing of sterile APIs are not covered by this guidance, but should be performed in accordance with GMP guidelines for drug (medicinal) products as defined by local authorities.

This Guide covers APIs that are manufactured by chemical synthesis, extraction, cell culture/fermentation, by recovery from natural sources, or by any combination of these processes. Specific guidance for APIs manufactured by cell culture/fermentation is described in Section 18.

This Guide excludes all vaccines, whole cells, whole blood and plasma, blood and plasma derivatives (plasma fractionation), and gene therapy APIs. However, it does include APIs that are produced using blood or plasma as raw materials. Note that cell substrates (mammalian, plant, insect or microbial cells, tissue or animal sources including transgenic animals) and early process steps may be subject to GMP but are not covered by this Guide. In addition, the Guide does not apply to medical gases, bulk-packaged drug (medicinal) products, and manufacturing/control aspects specific to radiopharmaceuticals.

Section 19 contains guidance that only applies to the manufacture of APIs used in the production of drug (medicinal) products specifically for clinical trials (investigational medicinal products).

An "API Starting Material" is a raw material, intermediate, or an API that is used in the production of an API and that is incorporated as a significant structural fragment into the structure of the API. An API Starting Material can be an article of commerce, a material purchased from one or more suppliers under contract or commercial agreement, or produced in-house. API Starting Materials normally have defined chemical properties and structure.

The company should designate and document the rationale for the point at which production of the API begins. For synthetic processes, this is known as the point at which "API Starting Materials" are entered into the process. For other processes (e.g., fermentation, extraction, purification), this rationale should be established on a case-by-case basis. Table 3-2 gives guidance on the point at which the API Starting Material is normally introduced into the process.

From this point on, appropriate GMP as defined in this Guide should be applied to these intermediate and/or API manufacturing steps. This would include the validation of critical process steps determined to impact the quality of the API. However, it should be noted that the fact that a company chooses to validate a process step does not necessarily define that step as critical.

The guidance in this document would normally be applied to the steps shown in gray in Table 3-2. It does not imply that all steps shown should be completed. The stringency of GMP in API manufacturing should increase as the process proceeds from early API steps to final steps, purification, and packaging. Physical processing of APIs, such as granulation, coating, or physical manipulation of particle size (e.g., milling, micronizing), should be conducted at least to the standards of this Guide.

This GMP Guide does not apply to steps prior to the introduction of the defined "API Starting Material.

Table 3-2 Application of this Guide to API manufacturing

Type of manufacturing		Application of this Guide to steps (shown in gray) used in this type of manufacturing			
Chemical manufacturing	Production of the API Starting Material	Introduction of the API Starting Material into process	Production of intermediate(s)	Isolation and purification	Physical processing and packaging
API derived from animal sources	Collection of organ, fluid, or tissue	Cutting, mixing, and/or initial processing	Introduction of the API Starting Material into process	Isolation and purification	Physical processing and packaging
API extracted from plant sources	Collection of plants	Cutting and initial extraction (s)	Introduction of the API Starting Material into process	Isolation and purification	Physical processing and packaging
Herbal extracts used as API	Collection of plants	Cutting and initial extraction		Further extraction	Physical processing and packaging
API consisting of comminuted or powdered herbs	Collection of plants and/or cultivation and harvesting	Cutting/ comminuting			Physical processing and packaging
Biotechnology: fermentation/ cell culture	Establishment of master cell bank and working cell bank	Maintenance of working cell bank	Cell culture and/ or fermentation	Isolation and purification	Physical processing and packaging
"Classical" fermentation to produce an API	Establishment of cell bank	Maintenance of the cell bank	Introduction of the cells into fermentation	Isolation and purification	Physical processing and packaging

Increasing GMP requirements →

2. Quality management

2.1 Principles

2.1.0 Quality should be the responsibility of all persons involved in manufacturing.

2.1.1 Each manufacturer should establish, document, and implement an effective system for managing quality that involves the active participation of management and appropriate manufacturing personnel.

2.1.2 The system for managing quality should encompass the organizational structure, procedures, processes, and resources, as well as activities necessary to ensure confidence that the API will meet its intended specifications for quality and purity. All quality-related activities should be defined and documented.

2.1.3 There should be a quality unit(s) that is independent of production and that fulfills both quality assurance (QA) and quality control (QC) responsibilities. This can be in the form of separate QA and QC units or a single individual or group, depending upon the size and structure of the organization.

2.1.4 The persons authorized to release intermediates and APIs should be specified.

2.1.5 All quality-related activities should be recorded at the time they are performed.

2.1.6 Any deviation from established procedures should be documented and explained. Critical deviations should be investigated, and the investigation and its conclusions should be documented.

2.1.7 No materials should be released or used before the satisfactory completion of evaluation by the quality unit(s) unless there are appropriate systems in place to allow for such use (e.g., release under quarantine as described in Section 10.20 or the use of raw materials or intermediates pending completion of evaluation).

2.1.8 Procedures should exist for notifying responsible management in a timely manner of regulatory inspections, serious GMP deficiencies, product defects, and related actions (e.g., quality-related complaints, recalls, regulatory actions, etc.).

2.2 Responsibilities of the quality unit(s)

2.2.0 The quality unit(s) should be involved in all qualityrelated matters.

2.2.1 The quality unit(s) should review and approve all appropriate quality-related documents.

2.2.2 The main responsibilities of the independent quality unit(s) should not be delegated. These responsibilities should be described in writing and should include but not necessarily be limited to:

i) Releasing or rejecting all APIs. Releasing or rejecting intermediates for use outside the control of the manufacturing company;

ii) Establishing a system to release or reject raw materials, intermediates, packaging, and labeling materials;

iii) Reviewing completed batch production and laboratory control records of critical process steps before release of the API for distribution;

iv) Making sure that critical deviations are investigated and resolved;

v) Approving all specifications and master production instructions;

vi) Approving all procedures impacting the quality of intermediates or APIs;

vii) Making sure that internal audits (self-inspections) are performed;

viii) Approving intermediate and API contract manufacturers;

ix) Approving changes that potentially impact intermediate or API quality;

x) Reviewing and approving validation protocols and reports;

xi) Making sure that quality-related complaints are investigated and resolved;

xii) Making sure that effective systems are used for maintaining and calibrating critical equipment;

xiii) Making sure that materials are appropriately tested and the results are reported;

xiv) Making sure that there is stability data to support retest or expiry dates and storage conditions on APIs and/or intermediates where appropriate; and

xv) Performing product quality reviews (as defined in Section 2.5).

2.3 Responsibility for production activities

The responsibility for production activities should be described in writing, and should include but not necessarily be limited to:

i) Preparing, reviewing, approving, and distributing the instructions for the production of intermediates or APIs according to written procedures.

ii) Producing APIs and, when appropriate, intermediates according to preapproved instructions.

iii) Reviewing all production batch records and ensuring that these are completed and signed.

iv) Making sure that all production deviations are reported and evaluated and that critical deviations are investigated and the conclusions are recorded.

v) Making sure that production facilities are clean and when appropriate disinfected.

vi) Making sure that the necessary calibrations are performed and records kept.

vii) Making sure that the premises and equipment are maintained and records kept.

viii) Making sure that validation protocols and reports are reviewed and approved.

ix) Evaluating proposed changes in product, process, or equipment.

x) Making sure that new and, when appropriate, modified facilities and equipment are qualified.

2.4 Internal audits (self-inspection)

2.4.0 In order to verify compliance with the principles of GMP for APIs, regular internal audits should be performed in accordance with an approved schedule.

2.4.1 Audit findings and corrective actions should be documented and brought to the attention of responsible management of the firm. Agreed corrective actions should be completed in a timely and effective manner.

2.5 Product quality review

2.5.0 Regular quality reviews of APIs should be conducted with the objective of verifying the consistency of the process. Such reviews should normally be conducted and documented annually and should include at least:

—A review of critical in-process control and critical API test results.

—A review of all batches that failed to meet established specification(s).

—A review of all critical deviations or nonconformances and related investigations.

—A review of any changes carried out to the processes or analytical methods;

—A review of results of the stability monitoring program;

—A review of all quality-related returns, complaints and recalls; and

—A review of adequacy of corrective actions.

2.5.1 The results of this review should be evaluated and an assessment made of whether corrective action or any revalidation should be undertaken. Reasons for such corrective action should be documented. Agreed corrective actions should be completed in a timely and effective manner.

引自：ICH Expert Working Group. ICH harmonised tripartite guideline: Good manufacturing practice guide for active pharmaceutical ingredients Q7[R], 2000.

Words and Expressions

guide	*n.* 指南	manufacturer	*n.* 制造商
pharmacopoeia	*n.* 药典	registration	*n.* 注册
filing	*n.* 备案	material	*n.* 原料
sterile	*adj.* 灭菌的	aseptic	*adj.* 无菌的
mammalian	*n.* 哺乳动物	substrate	*n.* 基质
plant	*n.* 植物	insect	*n.* 昆虫
transgenic	*adj.* 转基因	nonconformance	*n.* 不符合项
fragment	*n.* 片段	in-house	*adj.* 内部的
coating	*n.* 包衣	granulation	*n.* 造粒
deviation	*n.* 偏差	consistency	*n.* 一致性

Notes

① GMP: good manufacturing practice，药品生产质量管理规范。

② APIs: active pharmaceutical ingredients，药物活性成分。

③ QC: quality control，质量控制。

④ QA: quality assurance，质量保证。

⑤ cell culture: 细胞培养。

⑥ raw materials: 原料药。

⑦ chemical properties: 化学性质。

Exercises

1. Put the following into English:

一致性	临床试验	无菌的	成分
灭菌的	片段	疫苗	

2. Put the following into Chinese:

APIs	plasma	fermentation	cell culture
intermediate	pharmacopoeia	QA	QC

Unit 3
Isolation of Caffeine from Tea

In this experiment, caffeine will be isolated from tea leaves. The major problem of the isolation is that caffeine does not occur alone in tea leaves, but is accompanied by other natural substances from which it must be separated. The major component of tea leaves is cellulose, which is the major structural material of all plant cells. Cellulose is a polymer of glucose. Since cellulose is virtually insoluble in water, it presents no problems in the isolation procedure. Caffeine, on the other hand, is water soluble and is one of the major substances extracted into the solution called "tea". Caffeine comprises as much as 5 percent by weight of the leaf material in tea plants. Tannins also dissolve in the hot water used to extract tea leaves. The term tannin does not refer to a single homogeneous compound, or even to substances which have similar chemical structure. It refers to a class of compounds which have certain properties in common. Tannins are phenolic compounds having molecular weights between 500 and 3, 000. They are widely used to "tan" leather. They precipitate alkaloids and proteins from aqueous solutions. Tannins are usually divided into two classes: those which can be hydrolyzed and those which cannot. Tannins of the first type which are found in tea generally yield glucose and gallic acid when they are hydrolyzed. These tannins are esters of gallic acid and glucose. They represent structures in which some of the hydroxyl groups in glucose have been esterified by digalloyl groups. The non-hydrolyzable tannins found in tea are condensation polymers of catechin. These polymers are not uniform in structure, but catechin molecules are usually linked together at ring positions 4 and 8.

When tannins are extracted into hot water, the hydrolyzable ones are partially hydrolyzed, meaning that free gallic acid is also found in tea. The tannins, by virtue of their phenolic groups, and gallic acid by virtue of its carboxyl groups, are both acidic. If calcium carbonate, a base, is added to tea water, the calcium salts of these acids are formed. Caffeine can be extracted from the basic tea solution with chloroform, but the calcium salts of gallic acid and the tannins are not chloroform soluble and remain behind in the aqueous solution.

The brown color of a tea solution is due to flavonoid pigments and chlorophylls, as well as their respective oxidation products. Although chlorophylls are somewhat chloroform soluble, most of the other substances in tea are not. Thus, the chloroform extraction of the basic tea solution removes nearly pure caffeine. The chloroform is easily removed by distillation (b.p. 61℃) to leave the crude caffeine. The caffeine may be purified by recrystallization or by sublimation.

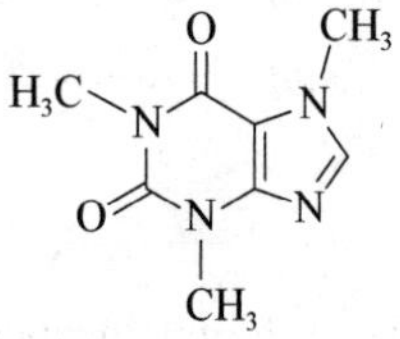

Caffeine

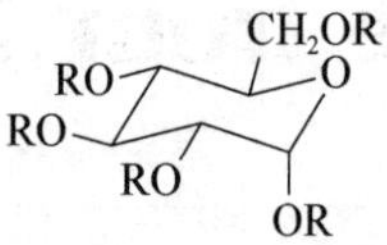

Glucose if R=H
A Tannin if some R=Digalloyl

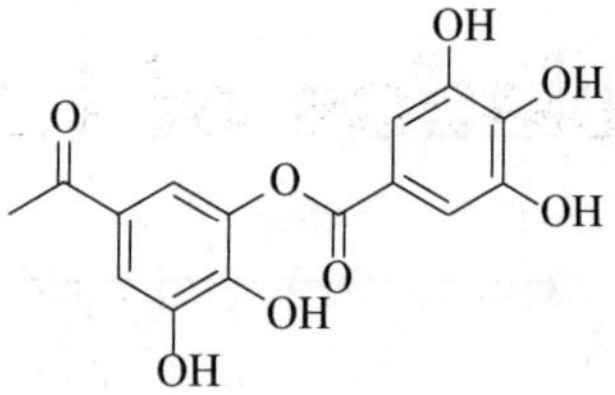

A Digalloyl Group

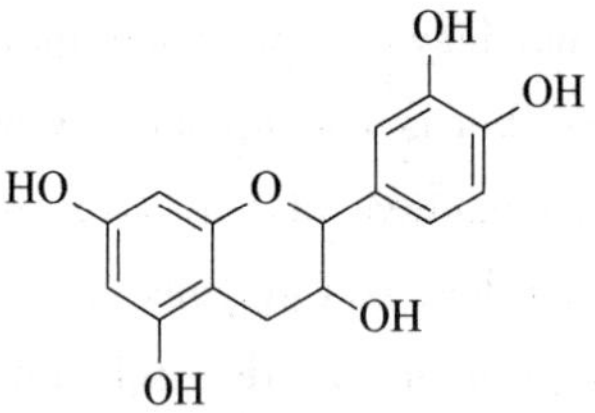

Catechin

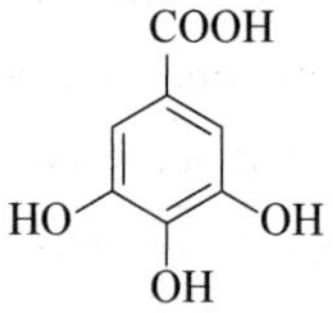

Gallic Acid

In a second part of this experiment, caffeine will be converted to a derivative. A derivative of a compound is a second compound, of known melting point, formed from the original compound by a simple chemical reaction. In trying to make a positive identification of an organic compound, it is often customary to convert it into a derivative. If the first compound, caffeine in this case, and its derivative both have melting points which match those reported in the chemical literature (e.g., a handbook), it is assumed that there is no coincidence and that the identity of the first compound, caffeine, has been definitely established.

Caffeine is a base and will react with an acid to give a salt. Using salicylic acid, a derivative salt of caffeine, caffeine salicylate, will be made in order to establish the identity of the caffeine isolated from tea leaves.

Special instructions: Be careful when handling chloroform. It is a toxic solvent, and you should not breathe it excessively or spill it on yourself. When discarding spent tea leaves, do not put them in the sink because they will clog the drain. Dispose of them in a waste container.

Procedure

Place 25 g of dry tea leaves, 25 g of calcium carbonate powder, and 250 mL of water in a 500 mL three neck round bottom flask equipped with a condenser for reflux. Stopper the unused openings in the flask and heat the mixture under reflux for about 20 minutes. Use a Bunsen burner to heat. While the solution is still hot, filter it by gravity through a fluted filter using a fast filter paper such as E&D No. 617 or S&S No. 595. You may need to change the filter paper if it clogs.

Cool the filtrate (filtered liquid) to room temperature and, using a separatory funnel, extract it twice with 25 mL portions of chloroform. Combine the two portions of chloroform in a 100 mL round bottom flask. Assemble an apparatus for simple distillation and remove the chloroform by distillation. Use a steam bath to heat. The residue in the distillation flask contains the caffeine and is purified as described below (crystallization). Save the chloroform that was distilled. You will use some of it in the next step. The remainder should be placed in a collection container.

Crystallization (Purification)

Dissolve the residue obtained from the chloroform extraction of the tea solution in about 10 mL of the chloroform that you saved from the distillation. It may be necessary to heat the mixture on a steam bath. Transfer the solution to a 50 mL beaker. Rinse the flask with an additional 5 mL of chloroform and combine this in the beaker. Evaporate the now light-green solution to dryness by heating it on a steam bath in the hood.

The residue obtained on evaporation of the chloroform is next crystallized by the mixed solvent method. Dissolve it in a small quantity (about 2 to 4 mL) of hot benzene and add just enough high boiling (60℃ to 90℃) petroleum ether (or ligroin) to turn the solution faintly cloudy. Alternatively, acetone may be used for simple crystallization without a second solvent. Cool the solution and collect the crystalline product by vacuum filtration using a Hirsch funnel. Crystallize the product the same way a second time if necessary, and allow the product to dry by allowing it to stand in the suction funnel for a while. Weigh the product. Calculate the weighty percentage yield based on tea and determine the melting point. If desired, the product may be further purified by sublimation as described in the next experiment.

The derivative

Dissolve 0.20 g of caffeine and 0.15 g of salicylic acid in 15 mL of benzene in a small beaker by warming the mixture on a steam bath. Add about 5 mL of high boiling (60℃ to 90℃) petroleum ether and allow the mixture to cool and crystallize. It may be necessary to cool the beaker in an ice water bath or to add a small amount of extra petroleum ether to induce crystallization. Collect the crystalline product by vacuum filtration using a Hirsch funnel. Dry the product by allowing it to stand in the air, and determine its melting point. Check the value against that in the literature. Submit the sample to the instructor in a labeled vial.

Words and Expressions

isolation	*n.* 分离	gallic acid	五倍子酸
hydroxyl group	羟基	phenolic	*adj.* 酚的
digalloyl group	鞣酰基	hydrolyze	*vi.* 水解
non-hydrolyzable	*adj.* 不可水解的	hydrolyzable	*adj.* 可水解的
carboxyl group	羧基	catechin	*n.* 儿茶酸
calcium carbonate	碳酸钙	esterify	*v.* (使)酯化

flavonoid pigment	黄酮类颜料	acidic	*adj.* 酸的,酸性的
salicylic acid	水杨酸	base	*n.* 碱
reflux	*n.* 回流	sublimation	*n.* 升华
stopper	*v.* 塞住;*n.* 塞子	Bunsen burner	本生灯
condenser	*n.* 冷凝器	separatory funnel	分液漏斗
chloroform	*n.* 氯仿	steam bath	蒸气浴
chlorophyll	*n.* 叶绿素	distillation flask	蒸馏瓶
distillation	*n.* 蒸馏	petroleum ether	石油醚
acetone	*n.*丙酮	Hirsch funnel	赫尔什漏斗
filter paper	滤纸	ice water bath	冰水浴
suction funnel	吸入漏斗	beaker	*n.* 烧杯
filtrate	*n.* 滤(出)液 *v.* 过滤	rinse	*v.* 冲洗,漂洗 *n.* 漂清,冲洗
extract	*vt.* 榨出,萃取,提取,蒸馏(出) *n.* 萃取物,提取物	precipitate	*vt.* 使沉淀; *vi.*沉淀; *n.* 沉淀物
tannin	*n.* 丹宁,丹宁酸,鞣酸	vial	*n.* 小瓶,小玻璃瓶; *vt.* 放……于小瓶中
homogeneous	*adj.* 均一的,均相的,均匀的,同质的	salicylate	*n.* 水杨酸盐,水杨酸酯
three neck round bottom flask	三口烧瓶	ligroin	*n.* 轻石油,石油英,粗汽油

Notes

① cellulose: 纤维素。D-葡萄糖以 β-1, 4-糖苷键连接而成的链状高分子,具有 $(C_6H_{10}O_5)_n$ 的组成,是维管束植物、地衣植物以及一部分藻类细胞壁的主要成分。

② alkaloid: 生物碱,亦称植物碱。它是含氮的碱性有机化合物,常以较小的量而对人或动物呈现显著的药理作用;多数具有吡啶、喹啉、异喹啉、吡咯烷、六氢吡啶、吲哚、托品烷、嘌呤等的环状结构;大部分为无色结晶性固体,在植物液泡内可与酸形成盐。

③ Since cellulose is virtually insoluble in water, it presents no problems in the isolation procedure.

副词主要作用是修饰动词、形容词或动词性/形容词性的短语。本句中 virtually 修饰形容词 insoluble。virtually 具有"实际上、几乎、差不多"等含义,根据上下文,其词义应为几乎,本句译为:由于纤维素几乎不溶于水,因此纤维素的分离很容易。

④ Caffeine comprises as much as 5 percent by weight of the leaf material in tea plants.

句中 A comprise as much as B percent by weight of C 是一个常用句型,用来表示 A 占 C 的重量百分比为 B。撰写英文科技论文、实验报告等时,可以套用。

⑤ The term tannin does not refer to a single homogeneous compound, or even to substances which have similar chemical structure.

句中 or even 很容易翻译错,误译为"乃至、以至"。通过对句子成分的分析可知,该句为并列句,后一个分句由 or 连接,即 The term tannin does not refer to a single homogeneous compound, or does not refer even to substances which have similar chemical structure. 英语并列句中,后一个分句常省略某些与前一个分句中相同的词,故英译汉时,为使语义清楚需增补所省略的词。

Exercises

1. Put the following into Chinese:

cellulose	glucose	chloroform	beaker
crystallization	purification	apparatus	filter paper
hydroxyl group	carboxyl group	benzene	acetone
evaporation	insoluble	condensation	residue

2. Put the following into English:

蒸馏	重结晶	升华	过滤
酸	碱	盐	水杨酸
水杨酸钙	水杨酸甲酯	冷凝器	塞子
分液漏斗	水解	可水解的	三口烧瓶

Reading material A:

2018 China-Japan-Korea Culture Exchange Forum Showcases Guizhou Tea Culture

"Tea is deeply rooted in the traditional culture of China, Japan and Korea. In East Asia, tea is not only a popular drink, but is regarded as a symbol of inner peace and humanity", Lin Yi, vice-president of the Chinese People's Association for Friendship with Foreign Countries, said at the 2018 China-Japan-Korea Culture Exchange Forum in Guiyang, Guizhou province on September 16.

With the theme "explore the potential of tea culture and build a community with a shared future for mankind", tea specialists from all three countries met to discuss the culture and history of the popular beverage.

Kuchong Chung, former president of Korean journal-dongA.com, said it was not difficult to seek common ground between the tea culture of China and Korea as tea was introduced by China to Korea in the Tang and Song Dynasties. Nowadays, Pu'er tea produced in Yunnan is considered a thoughtful gift in South Korea.

Miyasako Masaaki, honorary professor at the Tokyo University of Arts said tea culture helps reinforce East Asian cultural identity and community solidarity.

As foreign tea-lovers looked on, Hu Jicheng, vice director of Guizhou Province's Agriculture Committee and the Guizhou Provincial Tea Association, discussed how he hoped to promote Guizhou's tea brand to the world.

"Thanks to the local low temperature and cloudy climate, Guizhou has produced high quality tea since the Tang Dynasty. Today we have 7 million mu (4, 667 square kilometers) of tea planted here and hope more foreigners will enjoy the refreshing flavor and jade-green color of Guizhou tea," he explained.

引自:LI H Y. 2018 China-Japan-Korea Culture Exchange Forum showcases Guizhou tea culture[EB/OL]. (2018-09-27)[2022-07-22]. https://www.chinadaily.com.cn/m/guizhou/guiyang/2018-09/27/content_37008303.htm.

Reading material B:

Industrial Total Synthesis of Natural Medicines

Unit 4
Reactor Technology

Reactor technology comprises the underlying principles of chemical reaction engineering (CRE) and the practices used in their application. The focuses of reactor technology are reactor configurations, operating conditions, external operating environments, developmental history, industrial application, and evolutionary change. Reactor designs evolve from the pursuit of new products and uses, higher conversion, more favorable reaction selectivity, reduced fixed and operating costs, intrinsically safe operation, and environmentally acceptable processing.

Besides stoichiometry and kinetics, reactor technology includes requirements for introducing and removing reactants and products, efficiently supplying and withdrawing heat, accommodating phase changes and material transfers, assuring efficient contacting of reactants, and providing for catalyst replenishment or regeneration. Consideration must be given to physical properties of feed and products (vapor, liquid, solid, or combinations), characteristics of chemical reactions (reactant concentrations, paths and rates, operating conditions, and heat addition or removal), the nature of any catalyst used (activity, life, and physical form), and requirements for contacting reactants and removing products (flow characteristics, transport phenomena, mixing requirements, and separating mechanisms).

All the factors are interdependent and be considered together. Requirements for contacting reactants and removing products are a central focus in applying reactor technology; other factors usually are set by the original selection of the reacting system, intended levels of reactant conversion and product selectivity, and economic and environmental considerations.

Reactor types and characteristics

Specific reactor characteristics depend on the particular use of the reactor as a laboratory, pilot plant, or industrial unit. All reactors have in common selected characteristics of four basic reactor types: the well-stirred batch reactor, the semibatch reactor, the continuous-flow stirred-tank reactor, and the tubular reactor (Figure 3-1).

Batch reactor

A batch reactor is one in which a feed material is treated as a whole for a fixed period of time. Batch reactors may be preferred for small-scale production of high priced products, particularly if many sequential operations are employed to obtain high product yields, e.g., a process requiring a complex cycle of temperature-pressure-reactant additions. Batch reactors also may be justified when multiple, low volume products are produced in the same equipment or when continuous flow is difficult, as it is with highly viscous or sticky solids-laden liquids, e.g., in the manufacture of polymer resins where molecular weight and product quality are markedly affected by increasing viscosity and heat removal demands. Because residence times can be more uniform in batch reactors, better yields and higher selectivity may be obtained than with continuous reactors. This advantage exists when undesired reaction products inhibit the reaction, side reactions are of lower order than that desired, or the product is an unstable or reactive intermediate.

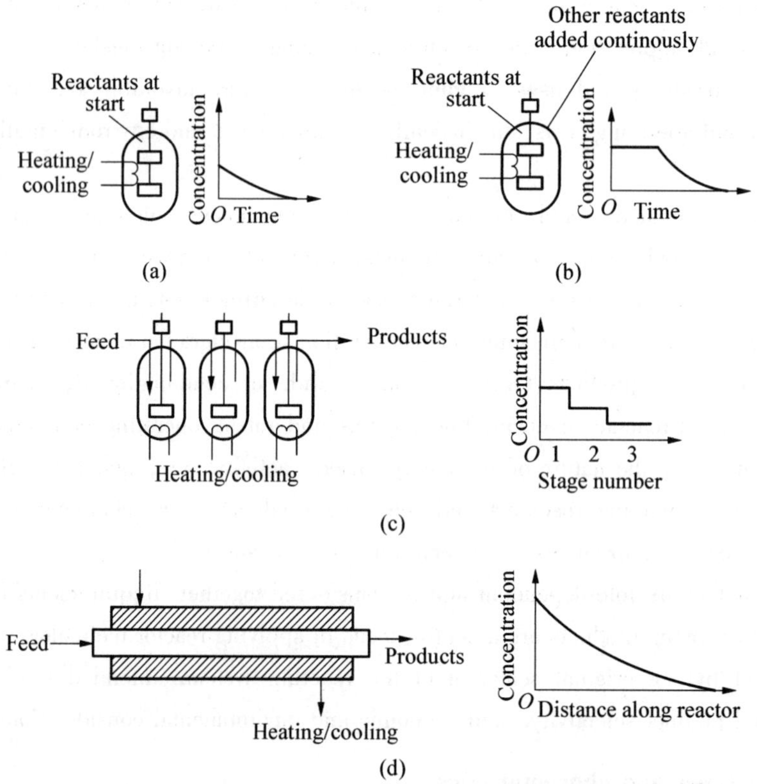

Figure 3-1 Reactor types: (a) batch, (b) semibatch, (c) continues-flow stirred-tank, and (d) tubular

Batch reactors often are used to develop continuous processes because of their suitability and convenient use in laboratory experimentation. Industrial practice generally favors processing continuously rather than in single batches, because overall investment and operating costs usually are less. Data obtained in batch reactors, except for very rapid reactions, can be well defined and used to predict performance of large scale, continuous-flow reactors. Almost all batch reactors are well stirred; thus, ideally, compositions are uniform throughout and residence times of all contained reactants are constant.

Semibatch reactor

The semibatch reactor is similar to the batch reactor but has the additional feature of continuous addition or removal of one or more components. For example, gradual addition of chlorine to a stirred vessel containing benzene and catalyst results in higher yields of di- and tri-chlorobenzene than the inclusion of chlorine in the original batch. Similarly, thermal decomposition of organic liquids is enhanced by continuously removing gaseous products. Constant pressure can be maintained and chain-terminating reaction products removed from the system. In addition to better yields and selectivity, gradual addition or removal assists in controlling temperature particularly when the net reaction is highly exothermic. Thus, use of a semibatch reactor intrinsically permits more stable and safer operation than in a batch operation.

Continuous-flow stirred-tank reactor

In a continuous-flow stirred-tank reactor (CSTR), reactants and products are continuously added and withdrawn. In practice, mechanical or hydraulic agitation is required to achieve uniform composition and temperature, a choice strongly influenced by process considerations, i.e., multiple specialty product requirements and mechanical seal pressure limitations. The CSTR is the idealized opposite of the well-stirred batch and tubular plug-flow reactors. Analysis of selected combinations of these reactor types can be useful in quantitatively evaluating more complex gas-, liquid-, and solid-flow behaviors.

Because the compositions of mixtures leaving a CSTR are those within the reactor, the reaction driving forces, usually reactant concentrations, are necessarily low. Therefore, except for zero- and negative-order reactions, a CSTR requires the largest volume of the reactor types to obtain desired conversions. However, the low driving force makes possible better control of rapid exothermic and endothermic reactions. When high conversions of reactants are needed, several CSTRs in series can be used. Equally good results can be obtained by dividing a single vessel into compartments while minimizing back-mixing and short-circuiting. The larger the number of stages, the closer performance approaches that of a tubular plug-flow reactor.

Tubular reactor

The tubular reactor is a vessel through which flow is continuous, usually at steady state, and configured so that conversion and other dependent variables are functions of position within the reactor rather than of time. In the ideal tubular reactor, the fluids flow as if they were solid plugs or pistons, and reaction time is the same for all flowing material at any given tube cross section; hence, position is analogous to time in the well-stirred batch reactor. Tubular reactors resemble batch reactors in providing initially high driving forces, which diminish as the reactions progress down the tubes.

Flow in tubular reactors can be laminar, as with viscous fluids in small-diameter tubes, and greatly deviate from ideal plug-flow behavior, or turbulent, as with gases, and consequently closer to the ideal. Turbulent flow generally is preferred to laminar flow, because mixing is introduced in the direction of flow. For slow reactions and especially in small laboratory and pilot-plant reactors, establishing turbulent flow can result in inconveniently long reactors or may require unacceptably high feed rates. Depending on the consequences in process development and impact on process economics, compromises, though necessary, may not prove acceptable.

Multiphase reactors

The overwhelming majority of industrial reactors are multiphase reactors. Some important reactor configurations are illustrated in Figure 3-2 and Figure 3-3. The names presented are often employed, but are not the only ones used. The presence of more than one phase, whether or not it is flowing, confounds analyses of reactors and increases the multiplicity of reactor configurations. Gases, liquids, and solids each flow in characteristic fashions, either dispersed in other phases or separately. Flow patterns in these reactors are complex and phases rarely exhibit idealized plug-flow or well-stirred flow behavior.

A fixed-bed reactor is packed with catalyst. If a single phase is flowing, the reactor can be analyzed as a tubular plug-flow reactor or modified to account for axial diffusion. If both liquid and gas or vapor are injected downward through the catalyst bed, or if substantial amounts of vapor are generated internally, the reactors are mixed-phase, downflow, and fixed-bed reactors. If the liquid and gas rates are so low that the liquid flows as a continuous film over the catalyst, the reactors are called trickle beds. Trickle beds have potential advantages of lower pressure drops and superior access for gaseous reactants to the catalyst; however restricted access can also be a disadvantage, e. g., where direct gas contact promotes undesired side reactions.

At higher total flow rates, particularly when the liquid is prone to foaming, the reactor is a pulsed column. This designation arises from the observation that the pressure drop within the catalyst bed cycles at a constant frequency as a result of liquid temporarily blocking gas or vapor pathways. The pulsed column is not to be confused with the pulse reactor used to obtain kinetic data in which a pulse of reactant is introduced into a tube containing a small amount of catalyst.

Downflow of reactants is preferred because reactors are more readily designed mechanically to hold a catalyst in place and are not prone to inadvertent excessive velocities, which upset the beds. Upflow is used less often but has the advantage of optimum contacting between gas, liquid, and catalyst over a wilder range of conditions. Mixed-phase, upflow, and fixed-bed reactors offer higher liquid holdups and greater assurance of attaining uniform catalyst wetting and radial flow distribution, the consequences of which are more uniform temperature distribution and greater heat transfer.

At high liquid flow rates in these co-current fixed-bed reactors, gas becomes the dispersed

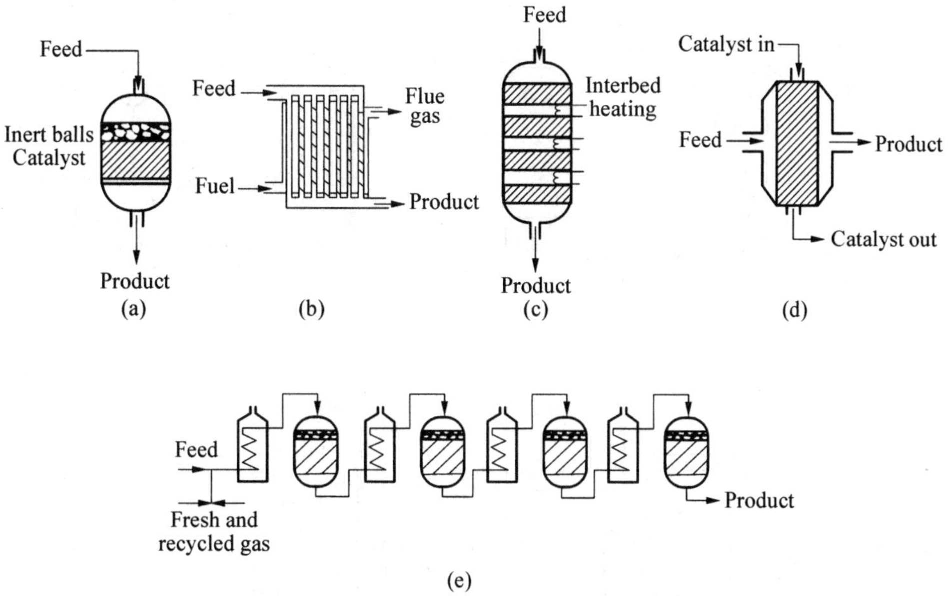

Figure 3-2 Multiple fixed-bed configurations: (a) adiabatic fixed-bed reactor, (b) tubular fixed beds, (c) staged adiabatic reactor with interbed heating (cooling), (d) moving radial fixed-bed reactor, and (e) trickle beds in series

phase and bubble flow develops; flow characteristics are similar to those in countercurrent packed-column absorbers. At high gas rates, spray and slug flows can develop. Moving beds are fixed-bed reactors in which spent catalyst or reactive solids are slowly removed from the bottom and fresh material is added at the top. A fixed bed that collects solids impurities present in the feed or produced in the early reaction stages is a guard bed. If catalyst deposits are periodically burned or otherwise removed, the operation is cyclic, and the catalyst remaining behind the combustion front is regenerated.

In bubble column reactors, gas bubble flow upward through a slower moving liquid. The bubbles, which rise in essentially plug flow, draw liquid in their wakes and thereby induce back-mixing in the liquid with which they have come in contact. Analogously, in spray columns, liquid as droplets descend through a fluid, usually gas. Both bubble and spray columns are used for reactions where high interfacial areas between phases are desirable. Bubble column reactors are used for reactions where the rate-limiting step is in the liquid phase, or for slow reactions where contacting is not critical. An important variant of the bubble column reactor is the loop reactor, commonly used for both multiphase and highly viscous systems. Loop reactors are distinguishable by their hydraulically or mechanically driven fluid recirculation, which offers the benefits of the well-stirred behavior of CSTRs and high average reactant concentrations of tubular reactors.

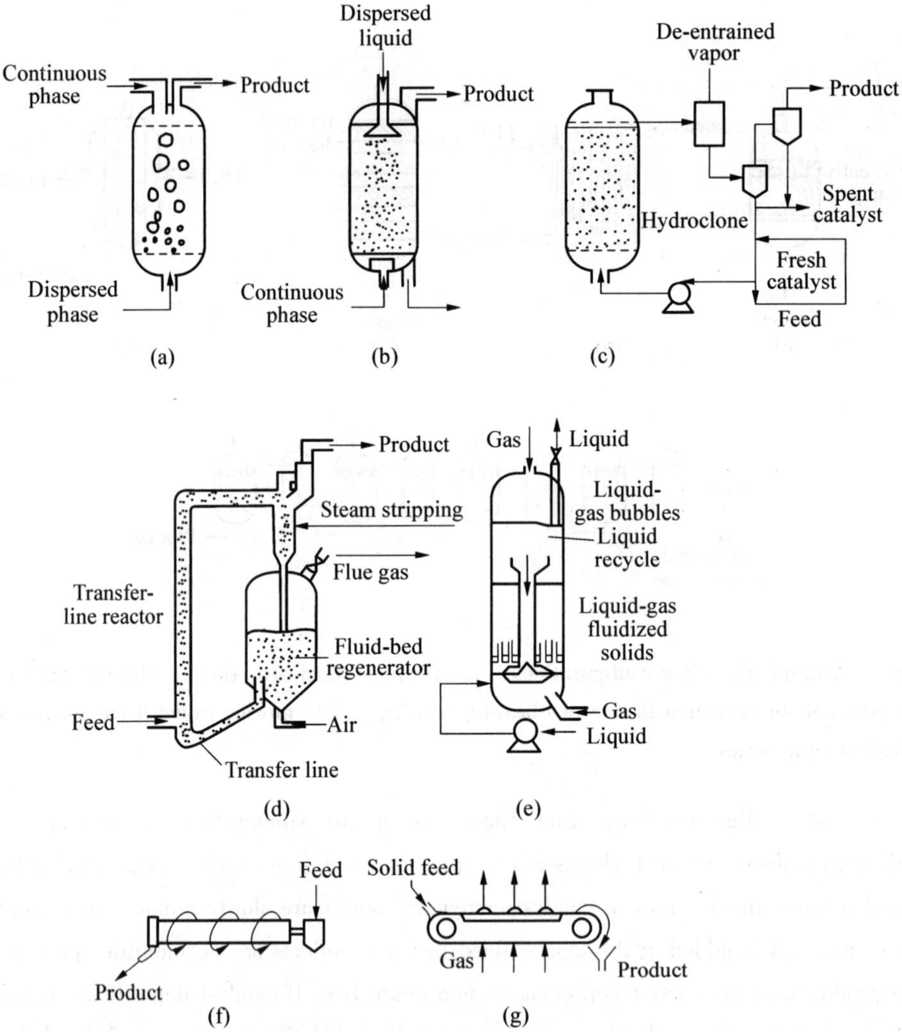

Figure 3-3 Multiphase fluid and fluid-solids reactors: (a) bubble column, (b) spray column, (c) slurry reactor and auxiliaries, (d) fluidization unit, (e) gas-liquid-solid fluidized reactor, (f) rotary kiln, and (g) traveling grate or belt drier

Reactors are termed fluidized or fluid beds if upward gas or liquid flows, alone or in concert, are sufficiently high to suspend the solids and make them appear to behave as a liquid. This process is usually referred to as fluidization. The most common fluid bed is the gas-fluidized bed. With gas feeds, the excess gas over the minimum required for fluidization rises as discrete bubbles, through which the surrounding solids circulate. At higher gas rates, such beds lose their clearly defined surface, and the particles are fully suspended. Depending on the circumstances, these reactors are variously called riser, circulating-fluidized, fast-fluidized, or entrainment reactors. In ebullating-bed or gas-liquid-solid reactors, the solids are fluidized by liquid and gas, with gas primarily providing lifting power in the former, and liquid in the latter.

These become slurry bubble column reactors (less precisely, slurry reactors) at high rates when the beds begin to lose their defined surfaces. Slurry bubble column reactors that contain finely powdered solids are often termed and treated as bubble column reactors because such suspensions are homogeneous.

A reactor is termed a radial or panel-bed reactor when gas or vapor flow perpendicular to a catalyst-filled annulus or panel. These are used for rapid reactions to reduce stresses on the catalyst or to minimize pressure drops. Similar cross-flow configurations also are used for processing solids moving downward under gravity while a gas passes horizontally through them. Rotary kilns, belt dryers, and travelling grates are examples. Cross-flow reactors are not restricted to solids-containing systems. Venturis, in which atomized liquids are injected across the gas stream, are effective for fast reactions and similarly for generating small gas bubbles in downward-flowing liquids where mass transport across the gas-liquid interface is limiting.

Words and Expressions

stoichiometry	*n.* 化学计算，化学计量(法，学)，理想配比法	exothermic	*adj.* 放热的，发热的
replenishment	*n.* (再)补给，充实，供给	endothermic	*adj.* 吸热的，内热的
pilot-plant	中试工厂，小规模试验性工厂	short circuit	短(捷)路，短接
batch	*adj.* 间歇的，分批的； *n.* 一批，批料，一次的分量	regime	*n.* 方式，方法，规范
tubular	*adj.* 管(状，形)的，筒式的	adiabatic	*adj.* 绝热的，不传热的
residence time	滞留时间	ebullating-bed	沸腾床
laminar	*adj.* 层(式，状，流)的，分层的，片(状)的	delineate	*vt.* 描外形，刻画，描绘
kiln	*n.*(砖，瓦)窑，(火)炉，干燥器； *vt.* (窑内)烘干，窑烧	entrainment	*n.* 挟带，夹带，带去，传输
radial	*adj.* 径向的，(沿)半径的放射的； *n.* 径向，光(射)线	slurry	*n.* 稀(泥，沙)浆；淤(砂)浆，悬浮体(液) *v.* 使变成泥浆
annulus	*n.* 环(带，形，状)物，环形套筒，环状空间		

Notes

① comprise: 可以是"包含和包括",也可以解释为"构成"。它可以表示一个整体由若干部分组成,也可以表示若干部分组成一个整体。是一个双向的动词,因此没有被动语态。例如:The committee comprises seven persons。委员会由七人组成。Nineteen articles comprise Book One. 十几篇文章组成了第一卷。

② compose: 表示若干部分组成一个整体,所以如果要表示一个整体由若干部分组成就要用被动语态了。如 Five elements compose this compound. 或 This compound is composed of five elements.

③ consist of: 表示一个整体由若干部分组成。由于 consist 是一个不及物动词,所以没有被动语态。如 Water consist of hydrogen and oxygen. 水由氢和氧组成。

Exercises

1. Put the following into English:

滞留时间	沸腾床	化学计算	短接	绝热的

2. Put the following into Chinese:

pilot-plant	annulus	entrainment	tubular
replenishment	liquid	homogeneous	fluidization

Reading material:

Ultrasound Assisted Continuous Processing in Microreactors with Focus on Crystallization and Chemical Synthesis: A Critical Review

Currently, most of the pharmaceutical industries rely on batch and semi-batch operations since there are fewer encrustation problems than continuous processes. Even then many issues like control of process parameters, or the reproducibility of the batch exist which needs to be addressed. To tackle these issues, continuous processes are being applied to yield benefits like smaller volumes, lower operational and labor expenses and greater product reproducibility. Application of continuous processing leads to advantages like efficient raw material usage, better control over process parameters, increased product quality and ease in scale-up of the process with less energy consumption. Even though the continuous processing offers many advantages, there is a difficulty in efficient design and operation often prompting the need for process intensification. Pharmaceutical industries especially are looking for the intensification of processes so as to obtain the higher efficiency in the manufacturing and quality control of pharmaceutical products. Based on the multidisciplinary and innovative strategies, process intensification (PI) offers compact equipment and intensified transport processes leading to more efficient processes. In recent years, PI has become a major area of research and has been receiving a significant consideration for the development of processes in chemical industries. Process intensification has resulted in many advances in chemical processes that have become milestones in this field and have had a huge impact on current processes being used in the chemical engineering and processing. Process intensification also leads to the technologies which are safer, cleaner, energy-efficient, and markedly smaller.

In the past few years, pharmaceuticals and fine chemical industries have attempted to include intensified processes based on using flow reactors (including those at micro level) for continuous processing of the end product and this has been one of the major recent revolutions in the field of process intensification. Microreactors are characterized with greater surface-to-volume ratios, better control over heat and mass transfer, controlled operating temperature, avoiding concentration irregularities, and better control over the process than the larger reactors. Even with these benefits, one major disadvantage limiting the use of microreactors is the problem of channel clogging for flows containing solid particles or where solid particles are likely to be formed, say for example crystallization operations or the precipitation based reactions. Moreover, higher fabrication cost and fixed design (i. e. there are very fewer opportunities to alter the design of microreactors) also add to the issues. Even then, there are many advantages of microreactors which make them a strong candidate for process intensification in continuous processes. The very physical constraints of microreactors lies in its lower production capacity while working with individual microreactor. Therefore, to increase mass throughput, scaling up of the microreactor is one of the options. Scaling up of microreactors is increasing the characteristic dimensions of microreactor for increased production but at the same time it would result in loss of advantage of intensification. The disadvantage can be effectively side-stepped to some degree through the scaling out of microfluidic systems. Instead of increasing the size of microreactors, scaling out simply indicates the increase of the number of microreactors in order to produce a parallel network. The benefits obtained in scaling out approach is that as the same size of the microreactors is used, it results in the same intensification in terms of chemistry performed at any level of scaling out. The scale up and process development in pharmaceutical industries ranges from sample production with few grams to kilograms sample production, and at pilot-plant scale, it is in the range of tons of pharmaceutical intermediates or active ingredients in a few weeks. For such cases, continuous flow microreactors allow for rapid experimental set-up of the laboratory plant for process design and optimization. When the microreactors are combined with the continuous processes, it benefits at laboratory as well at pilot scale.

The shortcoming of microreactors in terms of handling of solids needs to be sorted out as many chemical industries deal with solid catalysts, reactants, products or by-products. For instance, the share of final solid product produced by pharmaceuticals is about 80% and by fine chemicals is about 60% which is a considerable share by both of these industries. To overcome such problems, many approaches have been proposed such as modifying surface properties, application of multi-phase flow and application of the power ultrasound. Efforts have been also made to couple the microreactors with non-contact, non-classical, and sustainable energy sources. Out of these, combining the power ultrasound with microreactors is considered as a promising way. There are many reports on batch reactors and large-scale reactors where ultrasound has been effectively and efficiently applied for better mixing, improved heat and mass transfer. When low frequency ultrasound (US) is passed through microchannel, the cavitating bubbles resonance matches with the size of the channel making it beneficial to harness cavitation effect like intensified mixing, increased mass transfer, or the breakup of the agglomerates and also dislodge particles deposited or adhered on microchannel walls resulting in prevention in clogging. While talking about the higher frequency ultrasound application in microchannels, the wavelength matches the channel size, and forms the standing wave in the channel which upon formation give rise to the effects like acoustic radiation force and streaming adding synergies to microreactors.

Another constraint that limits the combination of microreactor with the ultrasound is that, due to the proximity to the channel walls, bubbles with a similar size as that of the channel could find it hard to collapse through cavitation. The static pressure inside the microreactor is high which has capacity to dissolve the gas bubbles formed in the microreactors. The bubble forming in the water needs to crossover the minimum negative pressure (Blake threshold pressure) of 1, 400 atm to overcome water molecule cohesion. But in the actual case, Blake threshold pressure of about 1 atm to 3 atm is sufficient to generate bubbles. The main reason for such lower values of pressure arises from the weakening spot generated for the molecular cohesion by the entrapped gas nuclei. These gas nuclei are generated by the small particles or defects present in the container holding the liquid. Once bubble is formed, it grows and goes through the different stages of cavitation depending on the resonance size. Even capillary waves generated by the ultrasound vibrations also goes through the cavitation events with sufficient forces provided by the interfacial tension of the liquid in microreactors.

Ultrasound (frequency: 20 kHz to several gigahertz) waves, when passed into a liquid medium at desired pressure intensities, generate cavitation which is responsible for process intensification. The mechanism of generation of cavitation and the different associated phases of cavitation are depicted in Figure 3-4. Low-frequency ultrasound is typically responsible for cavitation effects such as intensified mixing improved interfacial mass transfer, promoting de-agglomerates and dislocate particles deposited on surfaces to inhibit clogging. All these effects aid in achieving improved quality coupled with tailoring the desired product characteristics and also processes intensification. If resonance size of the induced cavitation bubble's matches the channel dimensions, it makes acoustic cavitation an ideal stage to harness its effects in the microreactors. In the case of high-frequency ultrasound, significant cavitation effects are normally not observed because of the operation below the cavitation threshold power. However, if the wavelength matches the channel size, it can form a standing wave in the channel that result into acoustic streaming and enhanced mixing. As mentioned earlier, on passing high frequency ultrasound through the medium, when the wavelength approaches that of the channel height or width, standing waves are formed in the microreactor. Due to the acoustic radiation forces, particle in the standing wave experiences the acoustophoresis and move towards the pressure node or antinode. Other than acoustophoresis, another phenomenon taking place with high frequency ultrasound is acoustic streaming, which is generally of two types, namely boundary layer streaming and Eckart streaming. Typically, Eckart streaming is observed over the order of a few centimetres and so these are not particularly found in the microreactors. Boundary layer streaming consisting of the Schlichting and Rayleigh streaming are majorly found in the microreactors. Rayleigh streaming, is result of the viscous dissipation of acoustic energy in the fluid boundary layer and is the main streaming phenomenon observed in microreactors.

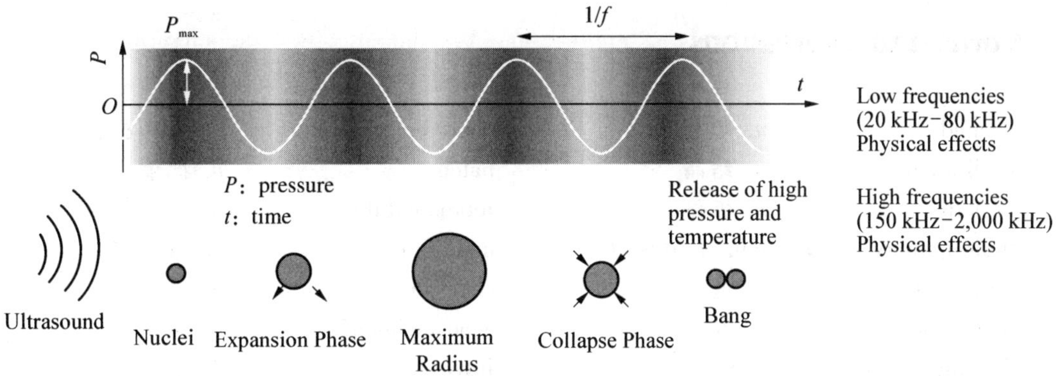

Figure 3-4 Mechanism of acoustic cavitation

Above discussed principles of cavitation and its effects have already been positively realized in microreactors for acousto-fluidic applications, such as cell/particle manipulation and fluid mixing for various chemical and biological systems. More recent studies also show that application of ultrasound over the range of 20 kHz–100 kHz frequency can reduce deposition of solid particles and prevent clogging. An important point to be noted is that it is very challenging for the ultrasound to be scaled-up to a commercial level as an individual operation. One study analyzed the scale up aspects of sonochemical reactors and illustrated the difficulties in scale up of ultrasonic reactors at industrial level. The major hurdles of scale up of ultrasonic reactors were very high degree of uncertainty due to larger scale up ratios, limitations over the proper reactor design depending on the experimental results and theoretical predictions about the bubble dynamics, issues related to the determination of cavity collapse intensity as a function of design and process parameters, wide distribution of energy dissipation patterns as well the concentrated zone of cavitation phenomena around the irradiation surface. Cavitation activity or efficiency of the sonochemical reactor is mainly dependent on the two major factors as the reactor geometry and operational ultrasound parameters. Therefore, for successful scaleup of the sonochemical reactors, it is important to understand the cavitational activity distribution and find newer designs and developments in the area of ultrasonic equipment coupled with their versatility and applicability in both batch as well as continuous flow processes. The current review primarily focuses on providing the state-of-the-art review on ultrasonic flow processes and advantages of their combination with microreactors. Positive effects of ultrasound in different microvolume continuous devices, such as continuous crystallizers, plug flow reactors, and channels are discussed. Important guidelines on the application of ultrasound in the fields of crystallization and chemical synthesis are presented in terms of best-operating conditions along with the discussion regarding the reactor designs with comments on suitability for scaleup.

引自：BANAKAR V V, SABNIS S S, GOGATE P R, et al. Ultrasound assisted continuous processing in microreactors with focus on crystallization and chemical synthesis: a critical review [J]. Chemical Engineering Research and Design, 2022, 182：273-289.

Words and Expressions

ultrasound	*n.* 超声	microreactor	*n.* 微反应器
crystallization	*n.* 结晶	batch	*n.* 批量
encrustation	*n.* 垢体	reproducibility	*n.* 可重复性
multidisciplinary	*adj.* 多学科的	microfluidic	*n.* 微流控
versatility	*n.* 通用性	cohesion	*n.* 内聚力
wavelength	*n.* 波长	acoustophoresis	*n.* 声泳
Schlichting	施利希廷	Rayleigh	瑞利
fluid	*n.* 流体	principle	*n.* 原则
cavitation	*n.* 空泡，成腔，空化	frequency	*n.* 频率
individual	*adj.* 个体的	sonochemical	*n.* 超声化学
clog	*v.* 堵塞，阻塞	process parameters	工艺参数

Notes

① fine chemical industries: 精细化工行业。

② pilot-plant scale: 中试规模。

③ the cavitation threshold power: 空化阈值功率。

④ process intensification: 过程强化。

⑤ Boundary layer streaming consisting of the Schlichting and Rayleigh streaming are majorly found in the microreactors.

在微反应器中主要存在由施利希廷流和瑞利流组成的边界层流。

Schlichting streaming: 习惯性指边界层内流。

Rayleigh streaming: 习惯性指边界层外流。

Exercises

1. Put the following into English:

结晶	可重复性	原则	多学科的	超声化学

2. Put the following into Chinese:

a.

ultrasound	cohesion	cavitation	process parameters
frequency	acoustophoresis	threshold	pharmaceutical

b. Important guidelines on the application of ultrasound in the fields of crystallization and chemical synthesis are presented in terms of best-operating conditions along with the discussion regarding the reactor designs with comments on suitability for scaleup.

Unit 5
European Pharmacopoeia: Imatinib Mesilate

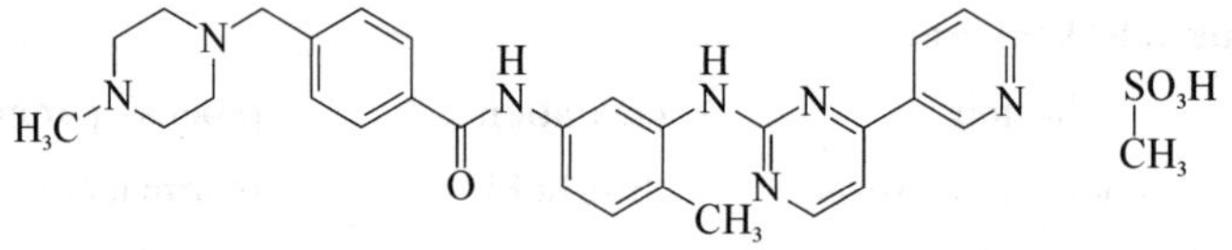

$C_{30}H_{35}N_7SO_4$　　$M_r = 589.7$

[220127-57-1]

Definition

4-[(4-Methylpiperazin-1-yl)methyl]-*N*-[4-methyl-3-[[4-(pyridin-3-yl)pyrimidin-2-yl]amino]phenyl]benzamide methanesulfonate.

Content: 98.0 per cent to 102.0 per cent (anhydrous substance).

Production

It is considered that alkyl methanesulfonate esters are genotoxic and are potential impurities in imatinib mesilate. The manufacturing process should be developed taking into consideration the principles of quality risk management, together with considerations of the quality of starting materials, process capability and validation. The general methods "2.5.37. Methyl, ethyl and isopropyl methanesulfonate in methanesulfonic acid", "2.5.38. Methyl, ethyl and isopropyl methanesulfonate in active substances" and "2.5.39. Methanesulfonyl chloride in methanesulfonic acid" are available to assist manufacturers.

Characters

Appearance: white or almost white, slightly brownish or yellowish powder; yellow or pale yellow, very hygroscopic, for the amorphous form.

Solubility: freely soluble in water, slightly soluble in ethanol (96 per cent), practically insoluble in methylene chloride.

It shows polymorphism (5.9).

Identification

Infrared absorption spectrophotometry (2.2.24).

Comparison: imatinib mesilate CRS.

If the spectra obtained in the solid state show differences, dissolve the substance to be examined and the reference substance separately in anhydrous ethanol R, evaporate to dryness and record new spectra using the residues.

Tests

Impurity F. Liquid chromatography (2.2.29) coupled with mass spectrometry (2.2.43).

Solvent mixture: acetonitrile R1, water for chromatography R (30: 70 *V/V*).

Test solution. Dissolve 50.0 mg of the substance to be examined in the solvent mixture and dilute to 100.0 mL with the solvent mixture.

Reference solution. Dissolve 2.0 mg of imatinib impurity F CRS in the solvent mixture and dilute to 100.0 mL with the solvent mixture. Dilute 1.0 mL of the solution to 200.0 mL with the solvent mixture. Dilute 1.0 mL of this solution to 10.0 mL with the solvent mixture.

Column:

—size: l = 0.15 m, Ø = 3.0 mm;

—stationary phase: end-capped octadecylsilyl amorphous organosilica polymer for chromatography R (3.5 μm);

—temperature: 40 ℃.

Mobile phase:

—mobile phase A: 1.26 g/L solution of ammonium formate R in water for chromatography R adjusted to pH 3.4–3.5 with anhydrous formic acid R;

—mobile phase B: 0.05 per cent V/V solution of anhydrous formic acid R in acetonitrile R1.

Table 3-3

Time (min)	Mobile phase A (per cent *V/V*)	Mobile phase B (per cent *V/V*)
0-6	80	20
6-10	80→20	20→80
10-15	20	80

NOTE: MS acquisition can be started at 3.5 min and stopped at 6 min; during non-acquisition the eluent is directed to waste.

Flow rate: 0.5 mL/min.

Detection: mass detector. The following settings have been found to be suitable and are given as examples; if the detector has different setting parameters, adjust the detector settings so as to comply with the system suitability criterion:

—ionisation: ESI-positive;

—detection m/z (SIM): 278.2;

—gas temperature: 350 ℃;

—drying gas flow: 12 L/min;

—nebuliser pressure: 414 kPa ;

—capillary voltage: 3 kV.

Injection: 10 μL.

System suitability: (reference solution):

—signal-to-noise ratio: minimum 20 for the principal peak;

—repeatability: maximum relative standard deviation of 10 per cent determined on 6 injections.

Calculation of percentage content:

—for impurity F, use the concentration of impurity F in the reference solution.

Limit:

—impurity F: maximum 20 ppm.

Impurity H. Liquid chromatography (2.2.29).

Solvent mixture: acetonitrile R1, water for chromatography R (30: 70 *V/V*).

Test solution. Dissolve 75.0 mg of the substance to be examined in the solvent mixture and dilute to 5.0 mL with the solvent mixture.

Reference solution (a). Dissolve the contents of a vial of imatinib impurity A CRS in 1.0 mL of the solvent mixture.

Reference solution (b). Dissolve 60.0 mg of imatinib impurity H CRS in the solvent mixture and dilute to 20.0 mL with the solvent mixture. Dilute 1.0 mL of the solution to 100.0 mL with the solvent mixture.

Reference solution (c). Dilute 5.0 mL of reference solution (b) to 50.0 mL with the solvent mixture.

Reference solution (d). Dissolve 0.150 g of the substance to be examined in the solvent mixture, add 1.0 mL each of reference solutions (a) and (b) and dilute to 10.0 mL with the solvent mixture.

Column:

—size: l=0.25 m, Ø=4.6 mm;

—stationary phase: end-capped octadecylsilyl silica gel for chromatography R (5 μm);

—temperature: 35 ℃.

Mobile phase:

—mobile phase A: dissolve 2.3 g of sodium octanesulfonate monohydrate R in 700 mL of water for chromatography R and add 300 mL of acetonitrile R1 and 1.2 mL of dilute phosphoric acid R;

—mobile phase B: dissolve 2.3 g of sodium octanesulfonate monohydrate R in 100 mL of water for chromatography R and add 900 mL of acetonitrile R1 and 1.2 mL of dilute phosphoric acid R;

Table 3-4

Time (min)	Mobile phase A (per cent V/V)	Mobile phase B (per cent V/V)
0—6	98	2
6—8	98→20	2→80
8—10	20	80

Flow rate: 2.3 mL/min.

Detection: spectrophotometer at 227 nm.

Injection: 10 μL of the test solution and reference solutions (c) and (d).

Identification of impurities: use the chromatogram obtained with reference solution (d) to identify the peaks due to impurities A and H.

Relative retention with reference to imatinib (retention time = about 8 min): impurity A = about 0.17; impurity H = about 0.2.

System suitability: reference solution (d):

—resolution: minimum 1.5 between the peaks due to impurities A and H.

Calculation of percentage content:

—for impurity H, use the concentration of impurity H in reference solution (c).

Limit:

—impurity H: maximum 0.02 per cent.

Related substances. Liquid chromatography (2.2.29).

Solvent mixture: acetonitrile R1, water for chromatography R (30:70 V/V).

Test solution. Dissolve 25.0 mg of the substance to be examined in the solvent mixture and dilute to 50.0 mL with the solvent mixture.

Reference solution (a). Dilute 1.0 mL of the test solution to 100.0 mL with the solvent mixture. Dilute 1.0 mL of this solution to 10.0 mL with the solvent mixture.

Reference solution (b). Dissolve 1 mg of imatinib for system suitability CRS (containing impurities A, B, C, D and J) in the solvent mixture and dilute to 2 mL with the solvent mixture.

Reference solution (c). Dissolve 25.0 mg of imatinib mesilate CRS in the solvent mixture and dilute to 50.0 mL with the solvent mixture.

Column:

—size: l = 0.25 m, Ø = 4.6 mm;

—stationary phase: end-capped octadecylsilyl silica gel for chromatography R (5 μm);

—temperature: 35 ℃.

Mobile phase:

—mobile phase A: dissolve 2.3 g of sodium octanesulfonate monohydrate R in 700 mL of water for chromatography R and add 300 mL of acetonitrile R1 and 1.2 mL of dilute phosphoric acid R;

—mobile phase B: dissolve 2.3 g of sodium octanesulfonate monohydrate R in 100 mL of water for chromatography R and add 900 mL of acetonitrile R1 and 1.2 mL of dilute phosphoric acid R;

Table 3-5

Time (min)	Mobile phase A (per cent *V/V*)	Mobile phase B (per cent *V/V*)
0-6	98	2
6-8	98→20	2→80
8-10	20	80

Flow rate: 2.3 mL/min.

Detection: spectrophotometer at 267 nm.

Injection: 10 μL of the test solution and reference solutions (a) and (b).

Identification of impurities: use the chromatogram supplied with imatinib for system suitability CRS and the chromatogram obtained with reference solution (b) to identify the peaks due to impurities A, B, C, D and J.

Relative retention with reference to imatinib (retention time = about 11 min): impurity A = about 0.2; impurity B = about 0.6; impurity J = about 0.9; impurity C = about 1.2; impurity D = about 2.3.

System suitability:

—resolution: minimum 3.0 between the peaks due to imatinib and impurity C in the chromatogram obtained with reference solution (b);

—signal-to-noise ratio: minimum 45 for the principal peak in the chromatogram obtained with reference solution (a);

—peak-to-valley ratio: minimum 1.3, where H_p = height above the baseline of the peak due to impurity J and H_v = height above the baseline of the lowest point of the curve separating this peak from the peak due to imatinib in the chromatogram obtained with reference solution (b).

Calculation of percentage contents:

—correction factors: multiply the peak areas of the following impurities by the corresponding correction factor: impurity A = 2.2; impurity B = 2.0;

—for each impurity, use the concentration of imatinib mesilate in reference solution (a).

Limits:

—impurity C: maximum 0.3 per cent;

—impurity D: maximum 0.2 per cent;

—impurities A, B: for each impurity, maximum 0.15 per cent;

—unspecified impurities: for each impurity, maximum 0.10 per cent;

—total: maximum 0.8 per cent;

—reporting threshold: 0.05 per cent.

Water (2.5.12): maximum 3.0 per cent, determined on 1.00 g.

Sulfated ash (2.4.14): maximum 0.1 per cent, determined on 1.0 g.

Assay

Liquid chromatography (2.2.29) as described in the test for related substances with the following modification.

Injection: test solution and reference solution (c).

Calculate the percentage content of $C_{30}H_{35}N_7SO_4$ taking into account the assigned content of imatinib mesilate CRS.

Impurities

Specified impurities: A, B, C, D, F, H.

Other detectable impurities [the following substances would, if present at a sufficient level, be detected by one or other of the tests in the monograph. They are limited by the general acceptance criterion for other/unspecified impurities and/or by the general monograph *Substances for pharmaceutical use* (2034). It is therefore not necessary to identify these impurities for demonstration of compliance. See also "5.10. Control of impurities in substances for pharmaceutical use"]: J.

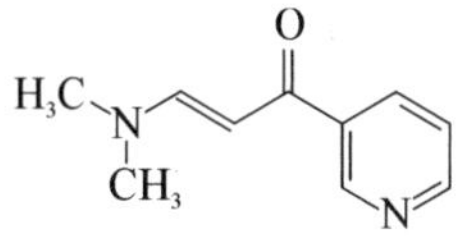

A. (2*E*)-3-(dimethylamino)-1-(pyridin-3-yl)prop-2-en-1-one,

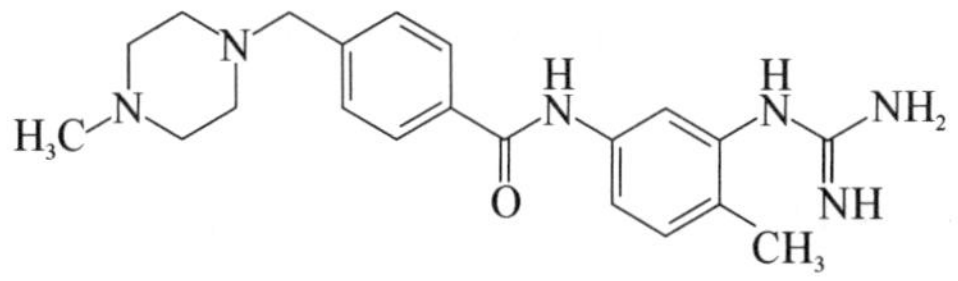

B. *N*-(3-carbamimidamido-4-methylphenyl)-4-[(4-methylpiperazin-1-yl) methyl] benzamide,

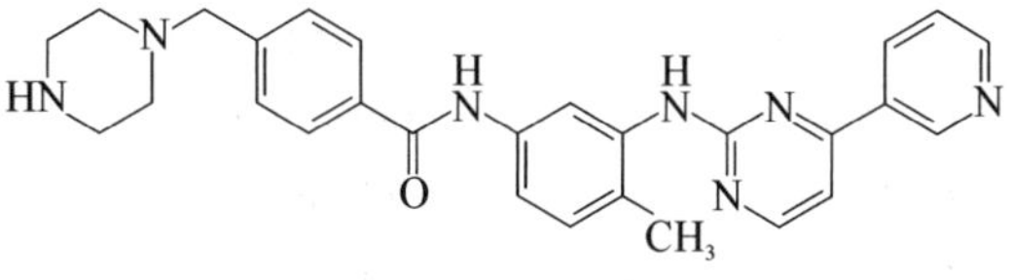

C. *N*-[4-methyl-3-[[4-(pyridin-3-yl) pyrimidin-2-yl] amino] phenyl]-4-(piperazin-1-yl methyl) benzamide (desmethylimatinib),

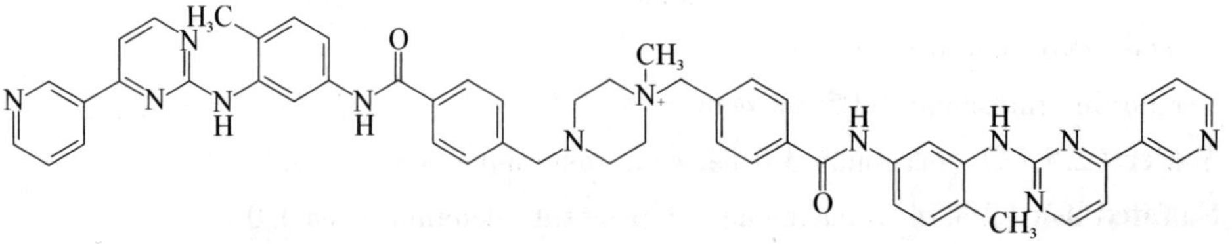

D. 1-methyl-1, 4-bis[4-[[4-methyl-3-[[4-(pyridin-3-yl)-pyrimidin-2-yl]amino]phenyl]carbamoyl]benzyl]-piperazin-1-ium (imatinib dimer),

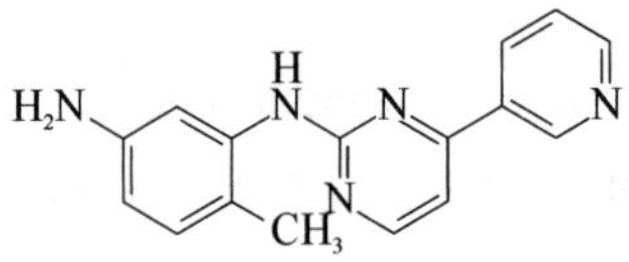

F. 6-methyl-*N*1-[4-(pyridin-3-yl)pyrimidin-2-yl]benzene-1, 3-diamine,

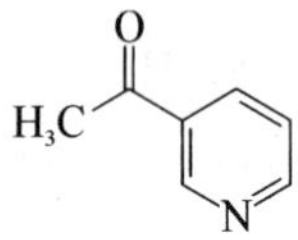

H. 1-(pyridin-3-yl)ethan-1-one,

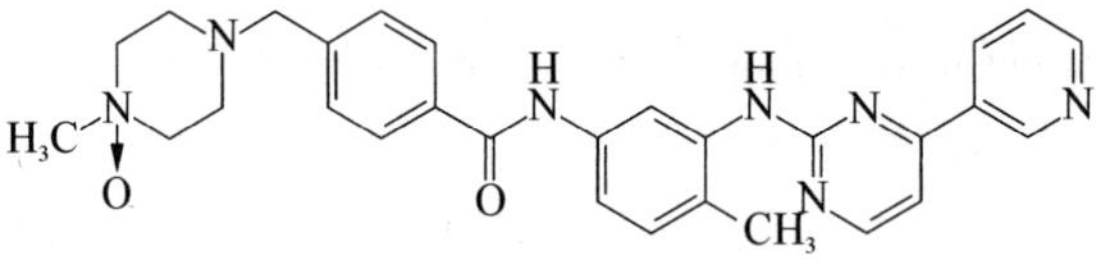

J. 4-[(4-methyl-4-oxidopiperazin-1-yl) methyl]-*N*-[4-methyl-3-[[4-(pyridin-3-yl)pyrimidin-2-yl]amino]phenyl]benzamide.

引自：European Directorate for the Quality of Medicine &Healthcare of the Council of Europe. European pharmacopoeia[M]. 10th ed. Strasbourg: Council of Europe, 2019.

Words and Expressions

mesilate	*n.* 甲磺酸	esters	*n.* 酯
alkyl	*n.* 烷基	genotoxic	*adj.* 遗传毒性
isopropyl	*n.* 异丙基	chloride	*n.* 氯化物
eluent	*n.* 洗脱液	ionisation	*n.* 电离
voltage	*n.* 电压	sodium	*n.* 钠
phosphoric	*adj.* 磷的	resolution	*n.* 分辨率
sulfate	*vi.* 硫酸化	formic acid	甲酸
Ammonium formate	甲酸铵		

Notes

① **Impurity F.** Liquid chromatography (2.2.29) coupled with mass spectrometry (2.2.43).
杂质 F. 液相色谱偶联质谱(液质联用)

② signal-to-noise ratio: 信噪比。

③ peak-to-valley ratio: 峰谷比。

Exercises

1. Put the following into English:

遗传毒性	洗脱液	硫酸化	氯化物

2. Put the following into Chinese:

ammonium formate	formic acid	mesilate	esters
retention time	chromatogram	Imatinib	starting materials

Chapter 4
Drugs Approval Applications and Others

Unit 1
Determining the Extent of Safety Data Collection Needed in Late Stage Premarket and Postapproval Clinical Investigations

1. Background

A robust safety database is critically important to a meaningful assessment of the potential risks of a new drug so that these can be weighed against its benefits. For this reason, extensive safety related data are collected throughout the course of drug development to characterize the safety profile of a drug. That drug profile continually evolves as the data accumulate from premarket testing through postmarket surveillance and risk management. However, in late stages of development or during the postmarket period, a selective, and better targeted, approach to safety data collection may be warranted. In some cases, for example, when certain aspects of the safety profile are well established prior to the completion of clinical development, it may no longer be necessary to collect certain types of safety data. Similarly, if there is generally a well established safety profile for a marketed drug being used in a postmarket clinical trial, it may not be necessary to collect certain types of safety data in such a trial.

In some cases, collection of data that are no longer useful for characterizing the safety profile of a drug may even have negative consequences. Arduous and excessive data collection may be a major disincentive to investigator participation in clinical trials. There is also growing interest in and a need for larger, simpler trials to obtain outcome data, data on long-term effects of drugs, and comparative effectiveness and safety data, but excessive data collection requirements may deter the conduct of these types of trials by increasing the difficulty, time, and cost of conducting the trials.

In such cases, more selective safety data collection may (1) improve the quality and utility of the safety database and safety assessment without compromising the integrity and validity of the trial results or losing important information, (2) ease the burden on investigators conducting and patients participating in a study, and (3) lower costs, thereby facilitating increased use of large, simple trials and better use of clinical trial resources generally. For these reasons, more selective safety data collection is generally advisable in the appropriate situations.

In the past, selective or specifically targeted data collection and reporting during clinical trials have been implemented on a case-by-case basis. However, there has been little public discussion of this practice. It is common, for example, in large outcome trials (generally conducted postmarket, but in some cases during late phase 3) for FDA to agree to selected or targeted collection of adverse events—specifically, those with important effects on treatment, such as events leading to discontinuation of therapy, change in dose, or need to add concomitant therapy — and more limited collection of routine laboratory data. It has also been common for FDA to agree to the collection of data only on concomitant therapy that was of a particular concern (e.g., because of a pertinent pharmacologic effect or potential interaction), rather than data on all concomitant therapy. This more targeted approach to collection of safety data reflected the view that less serious, less severe, and more common events (e. g., headache, nausea) had been adequately evaluated and characterized in earlier data collection and that little would be gained from further collection of these events. However, it is important to note that data from serious adverse events must always be collected and reported.

Encouraging more selective or targeted safety data collection in certain clinical trials and postmarket safety evaluations is also consistent with FDA's evolving overall approach to safety assessment, which emphasizes quality over quantity. For example, the recently revised regulations for reporting of serious, unexpected, suspected adverse reactions to FDA and all investigators (the IND safety reporting rule) address a similar concern. FDA determined that the old regulations on IND safety reports needed to be revised because large numbers of 15-day safety reports submitted to FDA under the old regulations did not reflect true signals of adverse reactions, but rather were background events that occur commonly in the study population, or study endpoints, and not interpretable as single events. Therefore, FDA modified the IND safety reporting regulations to require that single events, uninterpretable in isolation, not be reported, but be examined in the aggregate as safety data accumulate and be reported only when there are sufficient data to determine there is a reasonable possibility that the drug caused the adverse event.

The goal of this guidance is to provide advice on how and when to simplify data collection to maintain a balance between eliminating the collection of data that will not be useful and collecting sufficient data to allow adequate characterization of a drug's safety profile given the potential benefits. A sponsor considering a simplified data-collection approach should consult with the relevant FDA review division on the feasibility and acceptability of the plan before its implementation.

2. Targeted safety data collection—recommendations

The amount and types of safety data collected during clinical trials and observational safety evaluations will vary based on a range of factors, including the disease, patient population, subgroup of interest, preclinical findings, prior experience with the drug, experience with the drug class, phase of development, and study design, among other factors. Safety data collected could include some or all of the following, among other information:

- Serious, unexpected adverse events
- Serious expected adverse events
- Adverse events that cause discontinuation of treatment or dose modification

- Nonserious, unexpected adverse events
- Nonserious, expected adverse events
- Routine laboratory data (basic blood and urine analyses)
- Laboratory data limited to specific tests of interest or specialized laboratory data (e.g., radiographic tests, EKGs, pulmonary function tests, lipid fractions)
- Physical examination findings
- New concomitant medications (post-enrollment)

Certain aspects of patient history may also be relevant to safety, including:

- Concomitant illnesses at baseline
- Concomitant medications at baseline
- Prior treatment history
- Prior medical history
- Cardiovascular risk factors (history or laboratory)

Information on serious adverse events should generally be collected during all clinical trials, even for expected serious events, as collection of these events may change conclusions about the rate of occurrence or severity of an event. Adverse events leading to discontinuation or dose modification are also of interest, as such outcomes could vary by indication and population. Laboratory tests of particular interest for the study or population should also be collected. As discussed in more detail below, it may be appropriate in later phases of drug development or in postmarket studies to not collect some or all of the remaining types of data identified, or to collect such information at lower frequency or from only a fraction of the total study enrollment (e.g., 10% in a study with large population), depending on the extent to which a drug's safety characterization is already established.

2.1 When comprehensive data collection is generally needed

Although a more focused collection of data is appropriate in some circumstances and stages of development, there are circumstances when comprehensive data collection would be expected.

a. Development programs for original applications

In general, comprehensive data collection is expected throughout premarket clinical development for trials intended to support an original NDA or BLA approval. In these trials, it is important to collect sufficient data to determine the occurrence, dose-response, and subset (demographic, concomitant illnesses, concomitant therapy) variations for the full range of adverse events of the drug, not just serious events. However, even in a development program for a novel agent, there may be cases when sufficiently comprehensive safety data are available before the completion of clinical development to permit selective data collection.

b. Marketed drugs with differences in patient population, dose, or other conditions of use

Although it may be appropriate to use selective safety data collection in many postmarket trials, it is typically not appropriate when there are important differences in the patient population, dose, dosage regimen, duration of use, or route of administration compared to the conditions of use in the approved labeling. For example, comprehensive data collection would generally be needed in a development program for the prevention of cancer or for use as adjuvant therapy in the treatment of cancer for a drug that is already approved for the treatment of active cancer, including collection of important long-term events. The development program supporting the approval of the drug for treatment of existing cancer would have gathered information on the usually significant toxic effects of the drug in acute use, but would not have provided the information needed to support use of the drug in a longer duration adjuvant regimen in an essentially healthy population. Even in this setting, however, it may be acceptable to collect common, non-serious adverse events, such as headache, in only a subset of the overall study population.

c. Development programs for orphan indications

Orphan indications have limited patient populations available for study. Therefore, it is important in general to obtain comprehensive data on each patient to best inform product labeling.

d. Assessing specific adverse events and baseline risk factors

In some situations, the magnitude of the incidence or the relative risk of an adverse event is related to baseline factors, and it is useful to identify these factors to manage risk. In such situations, sufficient sample sizes will be needed to collect information adequate to characterize these risk factors.

2.2 Types of data that should typically be collected

There are also certain types of data that should almost always be collected:

a. Data in special populations

Because there is often limited availability of data in special populations, generally all data on use of a drug in special populations, including data on nonserious adverse events, should be collected. For example, pediatric patients often respond to a drug differently than adults. As a result, it is very important to collect all safety data in that population, especially as it is difficult to enroll pediatric patients in studies. There also is generally very limited information on exposure in pregnant and nursing women, so all safety data should be collected in those populations as well. Finally, there are often limited data available in the geriatric population, so these data should be collected as well.

b. Certain adverse event data

Data on all serious adverse events, deaths, and events that lead to dose modification or discontinuation should be collected at all times. In addition, data on adverse events that are troubling because of their potential seriousness, such as suicidality (suicidal ideation or thoughts), should generally be collected, unless existing data mitigate such a concern. Unscheduled visits, hospitalizations, and accidental injuries may also reflect adverse events of the drug, so these occurrences should also be collected. In an oncology setting, data from all Grade 3 and 4 adverse events, as well as Grade 2 events that affect vital organs (heart, liver) should always be collected.

c. Data for all study subject withdrawals

It is critical to collect data about and explore the reasons why a subject leaves a study. Designating withdrawals only as "withdrew consent" is uninformative and in most cases probably reflects inadequate pursuit of the underlying reason for withdrawal. A sponsor who discovers this designation in study data should take steps to educate investigators on the need to probe more diligently the reasons for dropouts. Narratives that more fully describe the circumstances for why subjects withdraw from treatment or declare that they cannot follow the protocol or dosing strategies may be very informative and should be collected along with adverse events occurring at those times. Such information may inform the cause-specific reasons for why some subjects remain in a study or remain on assigned treatment and others do not.

d. Targeted adverse event data

Data on adverse events that have been designated for reporting, based on the existing safety profile, pharmacology of the drug, or patient population, are expected to be collected, even if they are not serious, fatal, dose-modifying, or cause for discontinuation. Events would presumably be so designated when they are important (e.g., are a cause of discontinuation) and when the study may shed light on their occurrence (e.g., effects of dose, blood levels, concomitant therapy, concomitant illness, or demographic factors).

e. Long-term exposure to chronic usage treatments—characterizing the time course of risk

For chronically used treatments indicated for chronic diseases, it is often important to know the time course of events and whether the event rate or risk changes over the duration of exposure. Usually, it is important to have a sufficient denominator of patients followed long enough to observe and estimate the time-dependent risk (e.g., every three or six months of continued exposure). In such cases, it is important to fully collect important serious event data as exposure progresses.

Words and Expressions

premarket	*n.* 上市前	postapproval	*n.* 批准后
robust	*adj.* 强健的，健全的	adverse event	不良反应
validity	*n.* 有效性	nausea	*n.* 反胃，恶心
integrity	*n.* 完整性	dose modification	剂量调整
urine	*n.* 尿液	radiographic tests	(X)射线检测
pulmonary	*adj.* 肺的	prior medical history	既往病史
lipid	*n.* 脂质	prior treatment history	既往治疗史

demographic	*adj.* 人口统计学的	physical examination	体格检查
nursing	*n.*哺乳	cardiovascular	*adj.* 心血管的
pregnant	*adj.* 妊娠的,怀孕的	geriatric population	老年患者

Notes

① EKG: electrocardiogram, 心电图。也可缩写成 ECG。

② baseline: 基线期。Segen 医学词典给临床研究语境中基线(baseline)的解释是:基线是研究人群在前瞻性研究中最开始时的健康状况,是研究对象在接受试验组或对照组干预措施前的"0"时刻。药物的安全性和有效性可从基线数据的变化中评估,基线数据组间分布的差异或对结果评估造成偏倚。通常所谓的"基线"实则相对"随访"而言,专用于前瞻性研究设计,不过其他研究设计类型也可用"基线"泛指研究人群的基本情况。基线信息包括两方面的内容:(1) 研究人群的入选排除过程。先用入选标准粗略圈定分析人群,再用排除标准修正分析人群;(2) 研究人群基线特征的描述与比较。基线特征常常包括社会人口学特征、临床特征、实验室检查指标、疾病史和用药史等内容。

③ Grade 3 and 4 adverse events: 这里的分级指的是不良事件分级。分级系统采用美国国立癌症研究所的通用不良事件术语标准(CTCAE),美国健康与人类服务部门——美国国立卫生研究院、美国国立癌症研究院描述的分级系统用于癌症试验中的不良事件。4 级不良事件是导致需要救生紧急干预的事件;3 级不良事件可导致严重或临床重要但不会立刻危及生命的结果,如需要住院或住院期延长,伤残,导致洗澡、穿衣、脱衣、饮食、上厕所、服药能力受限;2 级事件导致需要中度、最低、局部或侵入性干预;导致相应年龄的做饭、在杂货店和/或服装店购买、使用电话、管理财物的能力受限。

④ "withdrew consent":撤回知情同意书。即受试者撤回参与研究的意愿,不再参与临床试验,是受试者随时退出研究权利的表现。

Exercises

1. Put the following into English:

上市后监督	研究终点	终止治疗	肺功能检查
给药方案	辅助治疗	血药水平	

2. Put the following into Chinese:

a. long-term effects of drug   prior experience with the drug
lipid fraction   inform product labeling.

b. Although it may be appropriate to use selective safety data collection in many postmarket trials, it is typically not appropriate when there are important differences in the patient population, dose, dosage regimen, duration of use, or route of administration compared to the conditions of use in the approved labeling.

Unit 2
Points to Consider on the Clinical Requirements of Modified Release Products Submitted as a Line Extension of an Existing Marketing Authorisation

Introduction

The development of modified-release preparations have a clinical rationale as it may reduce dose related side effects, improve efficacy and add to compliance to drug therapy.

Modified release products may be developed to reduced dose frequency which adds to convenience of use which in turn may facilitate compliance. Another rationale for developing modified release preparations is to smoothen the peaks of the plasma concentration curves in order to prevent peak concentration related adverse events.

Rarely a modified release preparation has been developed solely in order to mimic a TID or QID dosage schedule. In these cases the modified release preparation should be bioequivalent with the immediate release formulation given in dose schedule that is imitated.

In general, modified-release formulations are not bioequivalent to their immediate release form. Consequently, it might be difficult to assess whether the benefit/risk of the modified release is comparable to the corresponding doses of the immediately release form. Depending on the clinical setting additional clinical data will be required.

Types of modified release and dosage forms

Modified release dosage forms are formulations where the rate and/or site of release of the active ingredient(s) are different from that of the immediate release dosage form administered by the same route. This deliberate modification is achieved by special formulation design and/or manufacturing methods. Modified release dosage forms covered by this guideline include orally, intramuscularly, subcutaneously administered modified release and transdermal dosage forms.

- Prolonged release dosage forms: Prolonged release dosage forms are modified release dosage forms showing a sustained release compared to that of an immediate release dosage form administered by the same route.

- Delayed release dosage form: The release of the active substance from such modified release dosage forms is delayed for a certain period after administration or application of the dosage. The subsequent release is similar to that of an immediate release dosage form.

- Multiphasic release dosage forms:

—Biphasic Release: The first phase of drug release is determined by a fast release dose fraction providing a therapeutic drug level shortly after administration. The second extended release phase provides the dose fraction required to maintain an effective therapeutic level for a prolonged period.

—Pulsatile Release: Pulsatile drug release is intended to deliver a burst of drug release at specific time intervals.

• Multiple-unit: A multiple unit dosage form contains a plurality of units e.g. pellets or beads each containing release controlling excipients, e.g. in a gelatine capsule or compressed in a tablet

• Single-unit: The single-unit dosage forms consist of only one unit, e.g. osmotic tablet.

• Intramuscular/ subcutaneous depot formulations: A depot injection is usually a subcutaneous or intramuscular product which releases its active compound continuously over a certain period of time. Subcutaneous depot formulations include implants.

• Transdermal drug delivery systems (TDDS): A TDDS or transdermal patch is a flexible pharmaceutical preparation of varying size containing one or more active substance(s) to be applied on the intact skin for systemic availability.

There are two main types of transdermal patch systems depending on how the drug substance is dispersed in other patch components:

—Matrix systems with drug release based on the diffusion of drug substance.

—Reservoir systems containing a specific liquid drug compartment and release is controlled by a membrane.

Principles

As a principle, additional comparative clinical data are needed for modified release products developed as a line extension of an existing marketing authorisation UNLESS a justification for not doing so is given and accepted.

As this is a line extension and therefore the efficacy and safety of the immediate release product is known, the major issue would be to demonstrate that the new formulation is as effective as the existing one. Additionally the benefits of the new formulation should be shown or justified.

Whether these pharmacodynamic/clinical studie(s) should show equivalence or non-inferiority as compared to the standard formulation depends on the direction of the effect or safety issue at stake. In case efficacy and safety are closely related (e.g. anti-arrhythmic agents) equivalence studies are needed for showing that the effect studied remains within the equivalence margins. If an effect is bi-directional (e.g. thrombolytics, insulines) equivalence trials are needed as well. If an effect is unidirectional a demonstration of non-inferiority might be sufficient.

The type of studies that are required depends on whether appropriate, dynamic endpoints can be defined, whether the relationship between the dynamic markers and clinical efficacy is known, whether assay sensitivity is guaranteed and whether a non-inferiority margin or equivalence margins can be defined.

Such equivalence and non-inferiority studies may include a placebo arm besides the immediate and modified release preparation. A placebo arm is mandatory if assay sensitivity of the trial cannot be guaranteed.

In addition, equivalence margins or non-inferiority margins have to defined and justified irrespective whether the endpoint is based on, pharmacodynamic measurement or clinical variable.

If for a modified release product an indication is claimed different from that of the immediate release formulation a clinical development plan in accordance with existing guidelines or the state of art is required.

Requirements

In principle, additional comparative clinical data are needed for modified release products developed as a line extension of an existing marketing authorisation.

The rationale for the development of a modified release product, the existence of a plasma concentration-effect relationship and the condition studied determines the extend of these clinical data.

An applicant should justify the clinical development plan of a modified release product, i.e.: justify the rationale of the modified release product.

Justify in case an equivalence or non-inferiority study is performed, the choice of equivalence studies versus non-inferiority study, the lack of a placebo arm, the choice of equivalence margins for the endpoints chosen and the choice of the endpoints itself in the light of this rationale.

Justify the absence of a comparative study, e.g. provide evidence of a well-established plasma concentration effect relationship (with respect to efficacy, safety or both) in the light of this rationale.

引自：(1) European Agency for the Evaluation of Medicinal Products. Points to consider on the clinical requirements of modified release products submitted as a line extension of an existing marketing authorization[EB/OL]. (2003-11-17)[2023-12-20]. https://www.ikev.org/haber/bioav/187503line%20ext%20to%20MR.pdf.

(2) European Medicines Agency. Guideline on the pharmacokinetic and clinical evaluation of modified release dosage forms[EB/OL]. (2020-11-20)[2023-12-20]. https://www.ema.europa.eu/en/documents/scientific-guideline/guideline-pharmacokinetic-and-clinical-evaluation-modified-release-dosage-forms_en.pdf.

Words and Expressions

arrhythmic	*adj.* 心律不齐的	marketing authorisation	上市许可
osmotic	*adj.* 渗透的；渗透性的	bioequivalent	*adj.* 生物等效的
margin	*n.* 界值	release product	缓释制剂
thrombolytic	*n.* 溶栓剂	subcutaneously	*adv.* 皮下地
transdermal	*adj.* 经皮的，透皮的	pulsatile release	脉冲释放
gelatine	*n.* 明胶	placebo arm	安慰剂对照组
matrix	*n.* 基体；基质	plasma concentration	血药浓度
intramuscularly	*adv.* 肌内（地），肌内（注射）		

Notes

① TID: ter in die 每日三次。

② QID: qualer in die 每日四次。

③ immediate release: 常释或普通剂型。虽然从字面直译应是速释，但是《美国药典》（USP）〈1151〉对 immediate release（IR）的定义为"observed when nodeliberate effort has been made to modify the drug substance release profile"，即没有故意修饰药物的释放，比如使用崩解剂或润滑剂的胶囊和片剂，也被认为是 immediate release 的。欧洲药典中对 immediate release（也叫 conventional release）的定义为"a release of active substance（s）that is not deliberately modified by aspecial formulation design and/or manufacturing method"，即未采用特殊处方设计和/或生产方法故意修饰的药物释放。《中华人民共和国药典》（2020 版）〈9013〉只有缓释、控释和迟释制剂的定义。根据国家药品监督管理局药品审评中心（Center for Drug Evaluation, CDE）发布的《以药动学参数为终点评价指标的化学药物仿制药人体生物等效性研究技术指导原则》，其中提到的口服常释制剂即对应美国药品食品管理局（Food and Drug Administration, FDA）指南"Bioequivalence Studies with Pharmacokinetic Endpoints for Drugs Submitted Under an ANDA"中的 immediate release。因此，immediate release 在国内对应的翻译是常释。而常释剂型里面只有满足 15 分钟内活性药物成分（active pharmaceutical ingredient, API）的溶出达到标示量的 85%以上才叫快速溶出。

Exercises

1. Put the following into English:

提高疗效	增加药物依从性	血药浓度曲线	透皮吸收剂型
明胶胶囊	抗心律失常药	药代动力学标志物	测量灵敏度

2. Put the following into Chinese:

a. active ingredient　prolonged release　delayed release
   osmotic tablet　non-inferiority

b. Biphasic release: The first phase of drug release is determined by a fast release dose fraction providing a therapeutic drug level shortly after administration. The second extended release phase provides the dose fraction required to maintain an effective therapeutic level for a prolonged period.

c. A TDDS or transdermal patch is a flexible pharmaceutical preparation of varying size containing one or more active substance(s) to be applied on the intact skin for systemic availability.

Reading material:

Digital Pill That 'talks' to Your Smartphone Approved for First Time

In a groundbreaking decision on Monday, the Food and Drug Administration approved a drug with a "digital ingestion tracking system", which senses when a pill is swallowed and sends the data to a smartphone.

The new drug/device combination product called Abilify MyCite is approved for the treatment of schizophrenia and mood disorders. The Abilify (aripiprazole) tablets come embedded with an Ingestible Event Marker (IEM) sensor, the size of a grain of sand, that sends information to a patch the patient wears. The patch then transmits the data to smartphones and online healthcare portals—which can be accessed by health care professionals and caregivers if the patient approves (Figure 4-1).

This new concept could mean significant advancements in treatment for many disorders—specifically psychiatric illnesses, which rely heavily on patients who must consistently take their medication if they are to achieve stability.

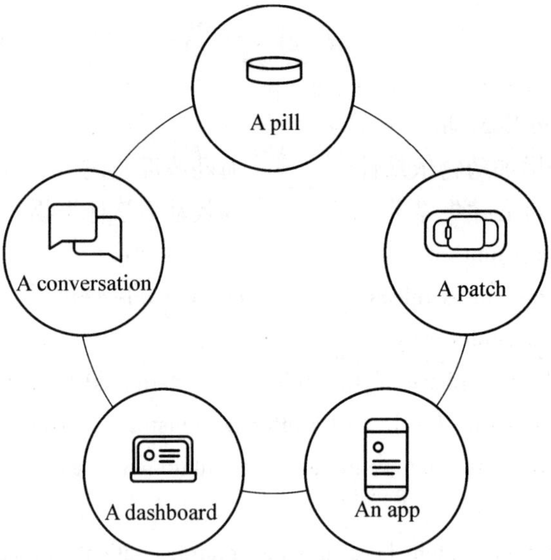

Figure 4-1 How ABILIFY MYCITE® Works

Schizophrenia, a chronic psychiatric disorder, affects approximately one percent of the U.S. population. The disabling condition includes delusions, thought disorders and hallucinations that can ruin jobs, relationships and day-to-day functions. People with schizophrenia have the greatest success in treatment if they take their medications.

This, however, can be difficult. If they don't take the medication, they relapse, are re-hospitalized and have to start over again. It's both disturbing to their lives and an extra expense for them to carry.

Psychiatric diseases are not the only illnesses that may benefit form a drug/device. This innovative way to track medications could help manage a variety of chronic illnesses.

Approximately half of all people with heart disease, for instance, don't take their medication regularly. For patients with diabetes, studies show that hospital costs are 41 percent higher every year for those who don't take their medication as directed, compared to those who do.

In general, medication "non-adherence" results in additional cost to the U.S. health care system of $ 290 billion annually.

While the potential benefits are easy to see, these pharmacological strides also raise concerns about diminished patient autonomy and suspicion of the medical system. Now, it's the patient who controls who has access to this electronic data. But this type of system could erode the trust traditionally shared between the physician and patient. That trust is particularly important for patients with psychiatric diseases.

"I would want a study conducted to see how the technology impacts the doctor-patient relationship, " Dr. Paul Applebaum, director of the Division of Law, Ethics and Psychiatry at Columbia University, said. "What kind of message are we communicating?"

It is also important to note that the improvement in patient compliance with treatment is the ideal outcome, but so far, nobody has shown that this pill will do it.

Applebaum sees how the new technology may prove beneficial in certain populations, such as "patients with early dementia, since the culprit for non-adherence is memory".

However, he notes it may not be as effective in patients who are worried about side effects or have other common reasons for not taking their prescriptions.

Sandy Walsh, a spokesperson for the FDA, comments "it is too soon to gauge" what implications this may have on a broader scale.

However, she points out that "the FDA supports the development and use of new technology in prescription drugs and is committed to working with companies to understand how technology might benefit patients and prescribers".

引自：ABC News. Digital pill that 'talks' to your smartphone approved for first time [EB/OL]. (2017-11-15)[2023-12-20]. https://abcnews.go.com/Health/digital-pill-talks-smartphone-approved-time/story?id=51161456.

Unit 3

USP Reference Standards

1. Introduction

Reference Standards provided by the United States Pharmacopeial Convention (USP Reference Standards or USP RS) are highly characterized materials demonstrated to have the appropriate qualities to support their intended use. USP RS are not for use in humans or animals.

USP RS are generally linked to relevant tests and assays in the *United States Pharmacopeia* (*USP*) or *National Formulary* (*NF*) documentary standards. They have been approved and established as suitable for use in the context of these applications. When approved as suitable for use in USP or NF tests and assays, USP RS also assume official status and legal recognition in the United States and other jurisdictions that recognize the USP or NF. Where USP or NF tests or assays call for the use of a USP RS, only those results obtained using the specified USP RS are conclusive.

USP RS may also be used to support other measurements not necessarily prescribed in USP-NF. Assessment of the suitability for use in other applications is the responsibility of the user.

2. Establishment approaches and value assignment

USP RS, when they are physical materials, are Reference Materials as defined in the *International Vocabulary of Metrology—Basic and General Concepts and Associated Terms* (*VIM*). The value assigned to quantitative standards relies on a mass determination by a primary reference measurement procedure such as a mass balance determination. If the value has been assigned by using a primary reference measurement procedure, the USP RS in metrology terms can be considered a primary measurement standard. If the value has been assigned by comparison or calibration to another material, the USP RS should be considered a secondary measurement standard; for example, when appropriate, USP RS are calibrated relative to international reference materials such as those provided by the World Health Organization (WHO).

USP may choose to develop USP RS that follow the requirements for the development of Certified Reference Materials (CRMs) in accordance with the relevant International

Organization for Standardization (ISO) Guides. Correct use of these CRMs support traceability of results to SI units and comparability of procedures. USP may also provide RS that are traceable to a CRM established by a National Metrology Institute or other official provider as required or referenced by the documentary standard.

USP may issue USP RS that are not required in a documentary standard. The material and performance attributes for these USP RS are described in the supporting documentation supplied with the USP RS.

USP RS go through a rigorous characterization as part of their establishment process. The types and extent of testing (including the number of laboratories) are primarily driven by the official uses of the standard, but material characterization goes beyond the establishment of suitability for use. Typically, a comparison to the previous lot is performed during the study when establishing a replacement lot of a USP RS as an additional verification of the suitability of the new standard lot.

3. USP Reference Standards for *USP* or *NF*

Official applications of USP RS are specified in USP or NF monographs and general chapters. These applications are as follows. Some USP RS could be used for several types of applications listed below.

3.1 Quantitative determinations

a. The majority of USP RS for quantitative determinations support measurements for total amounts of material on a mass basis. This category includes USP RS for USP or NF articles and impurity standards labeled for quantitative use. The assigned value of the USP RS is stated on the labeling and should be included in calculations used in the monograph and applicable general chapters.

b. USP RS for relative determinations of potency or activity are often required for the measurement of complex materials (e. g., biologics, antibiotics, herbals, some dietary supplements) and quantitative amounts may be expressed in units or relative potency terms other than mass. These USP RS are established by calibration to a primary standard where the property of the material determines the unit. For these standards where an International Standard (IS) established by the WHO exist, USP RS are National Measurement Standards. The USP RS documentation will indicate when the USP RS has been established by comparison to an International Standard (IS) established by the WHO. Results may be expressed in USP Units, Units, or International Units. Additional statements about unit/mass relationship, specific activity, or other relevant information related to the measurement may be provided in the USP RS documentation. For antibiotics that use microbial assays to determine activity, the potency is determined in units or micrograms per milligram (μg/mg) of activity. The units or μg/mg of activity is established against a WHO IS when one exists for that antibiotic. Where no WHO IS is presently available, USP establishes and maintains the standard to which USP RS lots are calibrated. This approach may also be chosen for other complex materials for which no WHO IS exists. In these instances, the USP standard is established in such a way as to ensure long-term stability and fitness for purpose, which permits the calibration of successive lots of USP RS with increased confidence that drift in the assigned unit can be avoided.

3.2 Qualitative determinations

a. Identification USP RS: USP RS for identification tests are typically presented as single components of high chemical purity, but may also be complex materials of natural, synthetic, or recombinant origin (e.g., biologics, natural products, botanicals, complex nonbiologicals, others). For complex materials, the identification attributes are presented in a matrix of other materials and require highly specific measurement systems (e.g., nucleic acid-based identity determination for naturally derived materials).

b. Impurity USP RS: Impurity USP RS are typically used for system suitability or as impurity markers. They may be presented as single-component materials, as mixtures containing more than one impurity, or as drug substance(s) containing one or more impurities.

c. Digital and Visual USP RS: Unlike chemical reference materials, these USP RS are not physical materials used in chemical analyses. Instead, these visual images are used by analysts to compare test articles to ensure that they meet compendial requirements.

3.3 Performance verification

These USP RS are typically called for in general tests and assays and are provided to analyze and, where appropriate, to facilitate adjustment of the operation of an instrument to ensure the results obtained are accurate and/or precise or otherwise give acceptable results. The use of these USP RS is generally described in associated general chapters and in the supporting documentation supplied with the USP RS.

4. USP Reference Standards for other measurements and determinations

USP also develops Reference Standards that may not be required in official *USP–NF* tests or assays. USP provides RS specified in the current edition of the *Food Chemicals Codex*, the *Herbal Medicines Compendium*, and standards referenced in regulatory requirements.

USP RS without an official use in the USP–NF are developed following the same quality systems used for the characterization and release of USP RS used in official tests and assays. These USP RS are generally intended to address common quality issues and challenges inherent to technologies that cut across different types of products (e.g., system suitability samples, calibrators used to demonstrate performance of an analytical procedure, process, or equipment). Extensive characterization of the USP RS candidate is required and the testing plan takes into account the use of different methods to measure the same attribute, demonstrating broader applicability of these standards. In the absence of a companion monograph or chapter, the information generated from these studies may be disseminated to the user via other types of supporting documents including but not limited to the USP Certificate.

5. Labeling

The labeling material consists of the label affixed to the USP RS and the associated USP Certificate. Both must be reviewed prior to handling or using the USP RS because in some cases not all of the necessary information can fit on the affixed label. USP Certificates are lot specific and are publicly available on the USP website (www.usp.org). Additional documentation may be provided with the USP RS as needed.

The affixed USP RS label typically contains the RS name, catalog number, lot number, package size, assigned value, storage conditions, handling instructions, and country of origin information. For multi-component items, there is also an outer package and label.

The affixed label also includes hazard and precautionary statements required by the Occupational Safety and Health Administration (OSHA) under the current revision of the Hazard Communication Standard (29 CFR 1910.1200). Terms used in these statements do not necessarily reflect specific definitions in the USP–NF. Safety Data Sheets for all USP RS are publicly available on the USP website.

In addition to the information provided on the affixed USP RS label, the USP Certificate will generally contain the RS chemical name and structure, sequence (if applicable), CAS number, molecular formula, and molecular weight. A typical chromatogram may also be included if necessary for the intended use. Additional information may be included such as special handling instructions or information needed for the use of the USP RS. The USP Certificate also includes a copy of the label text and a series of general instructions.

6. Packaging

The amount of material per individual USP RS container depends on the application of the standard. Some standards (mainly materials with significant handling requirements or materials that are available only in small amounts) are provided in single-use containers. Some single-use products may be lyophilized with content labeled in mass or activity units per container. If so labeled, the content of the container must be reconstituted in its entirety without any additional weighing. Instructions for use are given either on the label or USP Certificate, or in the monographs where the standard is used.

7. Storage

USP RS should be stored in the packaging configuration provided by USP, according to the label and USP Certificate instructions. When storage in refrigerator or freezer is stated on the label, follow the definitions given in *Packaging and Storage Requirements* <659>. If no specific directions or limitations are provided on the label, the conditions of storage shall be room temperature and protection from moisture, light, freezing, and excessive heat.

Any unused portions remaining after the container has been opened should be carefully stored in accordance with the user's Standard Operating Procedures and good laboratory practices. Decisions concerning the proper use of previously opened USP RS are the responsibility of the user, unless otherwise specified on the labeling. The user is responsible for ensuring that the contents of opened vials continue to be suitable for their intended use.

8. Continued suitability for use

All USP RS are periodically reevaluated by USP throughout their lifecycles. The USP Continued Suitability for Use (CSU) program is designed to monitor real-time suitability for use of all current lots of USP RS. Suitability testing intervals are established based on collaborative study data, manufacturer or supplier data, testing results, and CSU data trending and projections. When and where applicable, an accelerated degradation study may be performed to provide additional information on the stability of the USP RS and to support suitability testing intervals. The goal of the CSU program is to confirm the continued suitability of the material for use of a USP RS in its compendial applications during its valid use period.

9. Valid use date

USP RS lots are assigned a valid use date upon depletion. The valid use date is the last day upon which a particular lot of USP RS can be used. Typically, the valid use date assigned is one year from the date the last vial of a lot is sold.

It is the responsibility of the user to ascertain that a particular lot of a USP RS has official status either as a "Current Lot" or as a "Previous Lot" prior to the valid use date. Current and previous lot information, as well as the most current version of the catalog, can be found on the USP website.

10. Proper use

Many compendial tests and assays are based on comparison of a sample to a USP RS. In such cases, measurements are made on preparations of both the sample and the USP RS. Where it is directed that a standard solution or a standard preparation be prepared for a quantitative determination, it is intended that the USP RS substance be accurately weighed and subsequent dilutions be performed using volumetric apparatuses with, at least, the prescribed tolerances. Potential errors associated with the use of volumetric apparatus of small volume should be taken into account.

Whenever the labeled directions for use require either drying or a correction for water and/or volatiles, this should be performed at the time of use. Further experimental details should be controlled by the user's Standard Operating Procedures and good laboratory practices.

The following list of label terms and definitions is provided as guidance for the handling and use of USP RS:

- Assigned value (Calculation value)
- As is
- Anhydrous basis, determine water content titrimetrically at time of use
- Dried basis, determine loss on drying at time of use
- Dried material, dry before use, or use previously dried material

引自：The United States Pharmacopiecial Convention. The United States pharmacopeia: general chapters, DocID: 1_GUID-041BE2C7-30A1-44FA-AC03-6F6997D2F251_3_en-US. [S]. 2020.

Words and Expressions

lot	*n.* 批次	jurisdiction	*n.* 司法权，管辖权
metrology	*n.* 计量学	mass balance	物料平衡，质量守恒
calibrate	*v.* 校准	traceability	*n.* 追溯
chromatogram	*n.* 色谱	characterization	*n.* 表征
quantitative	*adj.* 定量的	antibiotic	*n.* 抗菌素，抗生素；*adj.* 抗生的，抗菌的
drift	*n.* 漂移	qualitative	*adj.* 定性的
compendial	*n.* 药典	reference standard	标准品
volatiles	*n.* 挥发物	dilution	*n.* 稀释，稀释法
hazard	*n.* 危害物	titrimetrically	*n.* 滴定法
lyophilize	*n.* 低压冻干法，升华干燥	botanical	*n.* 植物性药材；*adj.* 植物学的，来自植物的

Notes

① The United States Pharmacopeial Convention: 美国药典委员会。

② United States Pharmacopeia (USP) or National Formulary (NF):《美国药典/国家处方集》，简称(USP-NF)。1820 年，11 位医生在美国国会大厦的参议院商讨出版一部最佳治疗药品的汇编，给出适用的药名，并提供制剂的处方。1820 年 12 月 15 日，第一版《美国药典》出版，在该版的前言中提出，刊印该药典的目的是从具有治疗效力的物质中，选择那些功效充分证实的、作用明确的药物，并将其做成制剂使其效力得到最大的发挥。此外赋予各种药物一个确切而合适的名称，以防止医师与药师间的交流障碍与不确定性。随着时间的推移，USP 的性质从处方汇编改变为药品标准的汇编。出版周期也不断缩短，1840—1942 年每 10 年一版；1942—2000 年每 5 年一版；从 2002 年开始每年一版。1888 年，美国药学会出版了第一部国家处方集——《非法定制剂的国家处

方集》,简称 NF。1975 年 USP 与 NF 合并出版,称为 USP-NF。2020 年开始,《美国药典/国家处方集》(USP-NF)的发行从先前的“主本(Main Edition)”和“增补本(Supplements)”变更为“期(Issues)”,并且 USP-NF 以网络形式发布于 online.uspnf.com,不再保留整套药典的版本号(即不再以 USP39、USP44 等版本号命名)。因为在新的 USP-NF Online 平台中,USP 引入了以单个文件标准为中心的模型。无论单个文件标准是否正式生效,都不会再被关联至一个具体的药典版本,而是基于单独的文件标准本身。USP-NF online 中的每个单独文件都有其自己的正式生效日期(Official date)信息,它链接到 USP-NF Online 平台中的唯一永久性文件标识符(DocID)。该 DocID 不会因为每期出版而发生改变,除非该文件标准发生了修订。

③ International Vocabulary of Metrology—Basic and General Concepts and Associated Terms:国际计量学词汇——基础和通用概念及相关术语。

④ International Organization for Standardization: 国际标准化组织,简称为 ISO。ISO 成立于 1947 年,是全球最大最权威的国际标准化组织,该组织自我定义为非政府组织,官方语言是英语、法语和俄语。参加者包括各会员国的国家标准机构和主要工业、服务业企业,负责当今世界上绝大部分领域(包括军工、石油、船舶等垄断行业)的标准化活动。ISO 的宗旨是“在世界上促进标准化及其相关活动的发展,以便于商品和服务的国际交换,在智力、科学、技术和经济领域开展合作。”中国于 1978 年加入 ISO,2008 年正式成为 ISO 的常任理事国。

⑤ certified reference materials: 有证标准物质, 简称 CRM。附有证书的标准物质,其一种或多种特性值用建立了溯源性的程序确定,使之可溯源到准确复现的用于表示该特性值的计量单位,而且每个标准值都附有给定置信水平的不确定度。所有有证标准物质应符合《国际通用计量学基本术语》中给出的“计量基准标准”的定义。

⑥ SI: 源自法语 Système International d'Unités,国际单位制,是国际计量大会(CGPM)采纳和推荐的一种一贯单位制。SI 单位是国际单位制中由基本单位和导出单位构成一贯单位制的那些单位。在国际单位制中,将单位分成三类:基本单位、导出单位和辅助单位。

⑦ Food Chemicals Codex: 食品化学法典,简称 FCC。食品化学法典是国际公认的标准各论,用于验证食品成分的纯度及特性。

⑧ Herbal Medicines Compendium: 草药汇编,为美国药典委员会(USP)出版,是免费提供的在线资源,为草药中使用的草药成分提供标准。

⑨ Occupational Safety and Health Administration: 美国职业安全与健康管理局。隶属于美国劳工部。

⑩ CAS number: Chemical Abstracts Service (Registry) number, CAS(编)号又称 CAS 登录号。CAS 号是某种物质(化合物、高分子材料、生物序列、混合物等)的唯一的数字识别号码。CAS 为美国化学会的下设组织化学文摘社(Chemical Abstracts Service, 简称 CAS)。该社负责为每一种出现在文献中的物质分配一个 CAS 编号,这是为了避免化学物质有多种名称的麻烦,使数据库的检索更为方便。如今的化学数据库普遍都可以用 CAS 编号检索。

Exercises

1. Put the following into English:

定量测定 相对效力 长期稳定性 系统适用性

2. Put the following into Chinese:

a. standard lot USP monographs and general chapters

impurity standard qualitative determination

molecular formula accelerated degradation study

b. USP RS for relative determinations of potency or activity are often required for the measurement of complex materials (e. g., biologics, antibiotics, herbals, some dietary supplements) and quantitative amounts may be expressed in units or relative potency terms other than mass.

c. For antibiotics that use microbial assays to determine activity, the potency is determined in units or micrograms per milligram (μg/mg) of activity.

d. If no specific directions or limitations are provided on the label, the conditions of storage shall be room temperature and protection from moisture, light, freezing, and excessive heat.

Unit 4
Highlights of Prescribing Information for Gleevec—Imatinib Mesylate Tablet

1. Indications and usage

Gleevec is a kinase inhibitor indicated for the treatment of:

- Newly diagnosed adult and pediatric patients with Philadelphia chromosome positive chronic myeloid leukemia (Ph+ CML) in chronic phase. (1.1)
- Patients with Philadelphia chromosome positive chronic myeloid leukemia (Ph+ CML) in blast crisis (BC), accelerated phase (AP), or in chronic phase (CP) after failure of interferon-alpha therapy. (1.2)
- Adult patients with relapsed or refractory Philadelphia chromosome positive acute lymphoblastic leukemia (Ph+ ALL). (1.3)
- Pediatric patients with newly diagnosed Philadelphia chromosome positive acute lymphoblastic leukemia (Ph+ ALL) in combination with chemotherapy. (1.4)
- Adult patients with myelodysplastic/myeloproliferative diseases (MDS/MPD) associated with plateletderived growth factor receptor (PDGFR) gene re-arrangements. (1.5)
- Adult patients with aggressive systemic mastocytosis (ASM) without the D816V c-Kit mutation or with c-Kit mutational status unknown. (1.6)
- Adult patients with hypereosinophilic syndrome (HES) and/or chronic eosinophilic leukemia (CEL) who have the FIP1L1-PDGFRα fusion kinase (mutational analysis or fluorescence in situ hybridization [FISH] demonstration of CHIC2 allele deletion) and for patients with HES and/or CEL who are FIP1L1-PDGFRα fusion kinase negative or unknown. (1.7)
- Adult patients with unresectable, recurrent and/or metastatic dermatofibrosarcoma protuberans (DFSP). (1.8)
- Patients with Kit (CD117) positive unresectable and/or metastatic malignant gastrointestinal stromal tumors (GIST). (1.9)
- Adjuvant treatment of adult patients following resection of Kit (CD117) positive GIST. (1.10)

2. Dosage and administration

- Adults with Ph+ CML CP (2.2): 400 mg/day

- Adults with Ph+ CML AP or BC (2.2): 600 mg/day
- Pediatrics with Ph+ CML CP (2.3): 340 mg/m/day
- Adults with Ph+ ALL (2.4): 600 mg/day
- Pediatrics with Ph+ ALL (2.5): 340 mg/m/day
- Adults with MDS/MPD (2.6): 400 mg/day
- Adults with ASM (2.7): 100 mg/day or 400 mg/day
- Adults with HES/CEL (2.8): 100 mg/day or 400 mg/day
- Adults with DFSP (2.9): 800 mg/day
- Adults with metastatic and/or unresectable GIST (2.10): 400 mg/day
- Adjuvant treatment of adults with GIST (2.11): 400 mg/day
- Patients with mild to moderate hepatic impairment (2.12): 400 mg/day
- Patients with severe hepatic impairment (2.12): 300 mg/day

All doses of Gleevec should be taken with a meal and a large glass of water. Doses of 400 mg or 600 mg should be administered once daily, whereas a dose of 800 mg should be administered as 400 mg twice a day. Gleevec can be dissolved in water or apple juice for patients having difficulty swallowing. Daily dosing of 800 mg and above should be accomplished using the 400-mg tablet to reduce exposure to iron.

3. Dosage forms and strengths

Tablets (scored): 100 mg and 400 mg. (3)

4. Contraindications

None. (4)

5. Warnings and precautions

- Edema and severe fluid retention have occurred. Weigh patients regularly and manage unexpected rapid weight gain by drug interruption and diuretics. (5.1, 6.1)
- Cytopenia, particularly anemia, neutropenia, and thrombocytopenia, have occurred. Manage with dose reduction, dose interruption, or discontinuation of treatment. Perform complete blood counts weekly for the first month, biweekly for the second month, and periodically thereafter. (5.2)
- Severe congestive heart failure and left ventricular dysfunction have been reported, particularly in patients with comorbidities and risk factors. Monitor and treat patients with cardiac disease or risk factors for cardiac failure. (5.3)
- Severe hepatotoxicity, including fatalities may occur. Assess liver function before initiation of treatment and monthly thereafter or as clinically indicated. Monitor liver function when combined with chemotherapy known to be associated with liver dysfunction. (5.4)
- Grade 3/4 hemorrhage has been reported in clinical studies in patients with newly

diagnosed CML and with GIST. GI tumor sites may be the source of GI bleeds in GIST. (5.5)

- Gastrointestinal (GI) perforations, some fatal, have been reported. (5.6)
- Cardiogenic shock/left ventricular dysfunction has been associated with the initiation of Gleevec in patients with conditions associated with high eosinophil levels (e.g., HES, MDS/MPD, and ASM). (5.7)
- Bullous dermatologic reactions (e. g., erythema multiforme and Stevens-Johnson syndrome) have been reported with the use of Gleevec. (5.8)
- Hypothyroidism has been reported in thyroidectomy patients undergoing levothyroxine replacement. Closely monitor TSH levels in such patients. (5.9)
- Fetal harm can occur when administered to a pregnant woman. Apprise women of the potential harm to the fetus, and to avoid pregnancy when taking Gleevec. (5.10, 8.1)
- Growth retardation occurring in children and pre-adolescents receiving Gleevec has been reported. Close monitoring of growth in children under Gleevec treatment is recommended. (5.11, 6.2)
- Tumor Lysis Syndrome. Close monitoring is recommended. (5.12)
- Reports of motor vehicle accidents have been received in patients receiving Gleevec. Caution patients about driving a car or operating machinery. (5.13)
- Renal Toxicity. A decline in renal function may occur in patients receiving Gleevec. Evaluate renal function at baseline and during therapy, with attention to risk factors for renal dysfunction. (5.14)

6. Adverse reactions

The most frequently reported adverse reactions (greater than or equal to 30%) were edema, nausea, vomiting, muscle cramps, musculoskeletal pain, diarrhea, rash, fatigue, and abdominal pain. (6.1)

To report SUSPECTED ADVERSE REACTIONS, contact Novartis Pharmaceuticals Corporation at 1-888-669-6682 or FDA at 1-800-FDA-1088 or www.fda.gov/medwatch.

7. Drug interactions

- CYP3A4 inducers may decrease Gleevec C_{max} and area under curve (AUC). (2.12, 7.1, 12.3)
- CYP3A4 inhibitors may increase Gleevec C_{max} and AUC. (7.2, 12.3)
- Gleevec is an inhibitor of CYP3A4 and CYP2D6 which may increase the C and AUC of other drugs. (7.3, 7.4, 12.3)
- Patients who require anticoagulation should receive low-molecular weight or standard heparin and not warfarin. (7.3)

See 17 for Patient Counseling Information.

Revised: 3/2022

8. Part of the full prescribing information

8.1 Drug administration

The prescribed dose should be administered orally, with a meal and a large glass of water. Doses of 400 mg or 600 mg should be administered once daily, whereas a dose of 800 mg should be administered as 400 mg twice a day.

For patients unable to swallow the film-coated tablets, the tablets may be dispersed in a glass of water or apple juice. The required number of tablets should be placed in the appropriate volume of beverage (approximately 50 mL for a 100-mg tablet, and 200 mL for a 400-mg tablet) and stirred with a spoon. The suspension should be administered immediately after complete disintegration of the tablet(s).

For daily dosing of 800 mg and above, dosing should be accomplished using the 400-mg tablet to reduce exposure to iron.

Treatment may be continued as long as there is no evidence of progressive disease or unacceptable toxicity.

8.2 Dosage forms and strengths

- 100 mg film coated tablets

Very dark yellow to brownish orange, film-coated tablets, round, biconvex with bevelled edges, debossed with "NVR" on one side, and "SA" with score on the other side.

- 400 mg film coated tablets

Very dark yellow to brownish orange, film-coated tablets, ovaloid, biconvex with bevelled edges, debossed with "gleevec" on one side and score on the other side.

8.3 Description

Imatinib is a small molecule kinase inhibitor. Gleevec film-coated tablets contain imatinib mesylate equivalent to 100 mg or 400 mg of imatinib free base. Imatinib mesylate is designated chemically as 4-[(4-Methyl-1-piperazinyl)methyl]-*N*-[4-methyl-3-[[4-(3-pyridinyl)-2-pyrimidinyl]amino]-phenyl]benzamide methanesulfonate and its structural formula is shown in Figure 4-2.

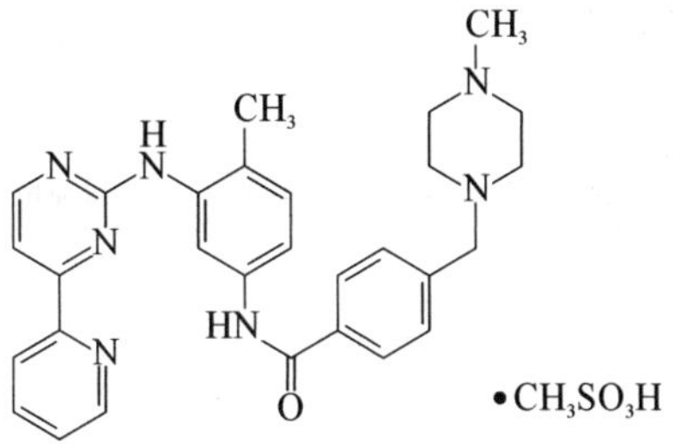

Figure 4-2 Structural formula of Imatinib Mesylate

Imatinib mesylate is a white to off-white to brownish or yellowish tinged crystalline powder. Its molecular formula is $C_{29}H_{31}N_7O \cdot CH_4SO_3$ and its molecular weight is 589.7 g/mol. Imatinib mesylate is soluble in aqueous buffers less than or equal to pH 5.5 but is very slightly soluble to insoluble in neutral/alkaline aqueous buffers. In non-aqueous solvents, the drug substance is freely soluble to very slightly soluble in dimethyl sulfoxide, methanol, and ethanol, but is insoluble in n-octanol, acetone, and acetonitrile.

Inactive Ingredients: colloidal silicon dioxide (NF); crospovidone (NF); hydroxypropyl methylcellulose (USP); magnesium stearate (NF); and microcrystalline cellulose (NF). Tablet coating: ferric oxide, red (NF); ferric oxide, yellow (NF); hydroxypropyl methylcellulose (USP); polyethylene glycol (NF), and talc (USP).

8.4 Principal display panel (Figure 4-3 and Figure 4-4)

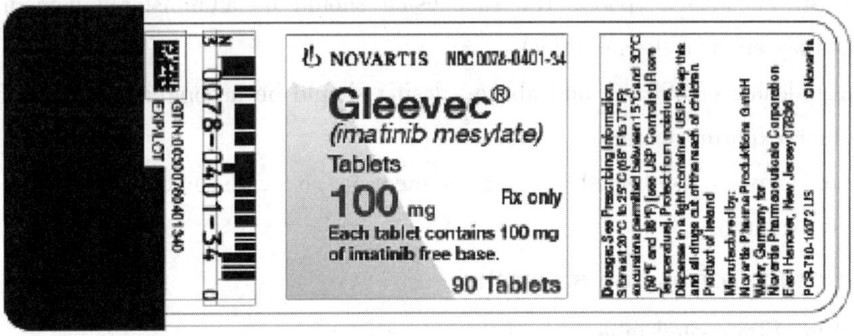

Figure 4-3 Principal display panel of Gleevec® (100 mg per tablet)

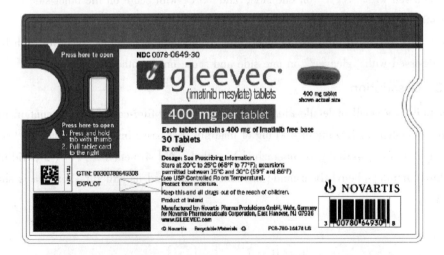

Figure 4-4 Principal display panel of Gleevec® (400 mg per tablet)

Supplementary Documentation

Full Prescribing Information: Contents *

1. Indications and Usage

1.1 Newly Diagnosed Philadelphia Positive Chronic Myeloid Leukemia (Ph+ CML)

1.2 Ph+ CML in Blast Crisis (BC), Accelerated Phase (AP) or Chronic Phase (CP) After Interferon-alpha (IFN) Therapy

1.3 Adult Patients with Ph+ Acute Lymphoblastic Leukemia (ALL)

1.4 Pediatric Patients with Ph+ Acute Lymphoblastic Leukemia (ALL)

1.5 Myelodysplastic/Myeloproliferative Diseases (MDS/MPD)

1.6 Aggressive Systemic Mastocytosis (ASM)

1.7 Hypereosinophilic Syndrome (HES) and/or Chronic Eosinophilic Leukemia(CEL)

1.8 Dermatofibrosarcoma Protuberans (DFSP)

1.9 Kit+ Gastrointestinal Stromal Tumors (GIST)

1.10 Adjuvant Treatment of GIST

2. Dosage and Administration

2.1 Drug Administration

2.2 Adult Patients with Ph+ CML CP, AP, or BC

2.3 Pediatric Patients with Ph+ CML CP

2.4 Adult Patients with Ph+ ALL

2.5 Pediatric Patients with Ph+ ALL

2.6 Adult Patients with MDS/MPD

2.7 Adult Patients with ASM

2.8 Adult Patients with HES/CEL

2.9 Adult Patients with DFSP

2.10 Adult Patients with Metastatic and/or Unresectable GIST

2.11 Adult Patients with Adjuvant GIST

2.12 Dose Modification Guidelines

2.13 Dose Adjustment for Hepatotoxicity and Non-Hematologic Adverse Reactions

2.14 Dose Adjustment for Hematologic Adverse Reactions

3. Dosage Forms and Strengths

4. Contraindications

5. Warnings and Precautions

5.1 Fluid Retention and Edema
5.2 Hematologic Toxicity
5.3 Congestive Heart Failure and Left Ventricular Dysfunction
5.4 Hepatotoxicity
5.5 Hemorrhage
5.6 Gastrointestinal Disorders
5.7 Hypereosinophilic Cardiac Toxicity
5.8 Dermatologic Toxicities
5.9 Hypothyroidism
5.10 Embryo-Fetal Toxicity
5.11 Growth Retardation in Children and Adolescents
5.12 Tumor Lysis Syndrome
5.13 Impairments Related to Driving and Using Machinery
5.14 Renal Toxicity

6. Adverse Reactions

6.1 Clinical Trials Experience
6.2 Postmarketing Experience

7. Drug Interactions

7.1 Agents Inducing CYP3A Metabolism
7.2 Agents Inhibiting CYP3A Metabolism
7.3 Interactions with Drugs Metabolized by CYP3A4
7.4 Interactions with Drugs Metabolized by CYP2D6

8. Use in Specific Populations

8.1 Pregnancy
8.2 Lactation
8.3 Females and Males of Reproductive Potential
8.4 Pediatric Use
8.5 Geriatric Use
8.6 Hepatic Impairment
8.7 Renal Impairment

10. Overdosage

11. Description

12. Clinical Pharmacology

12.1 Mechanism of Action
12.2 Pharmacokinetics

13. Nonclinical Toxicology

13.1 Carcinogenesis, Mutagenesis, Impairment of Fertility
13.2 Animal Toxicology and/or Pharmacology

14. Clinical Studies

14.1 Chronic Myeloid Leukemia
14.2 Pediatric CML
14.3 Acute Lymphoblastic Leukemia
14.4 Pediatric ALL
14.5 Myelodysplastic/Myeloproliferative Diseases
14.6 Aggressive Systemic Mastocytosis
14.7 Hypereosinophilic Syndrome/Chronic Eosinophilic Leukemia
14.8 Dermatofibrosarcoma Protuberans
14.9 Gastrointestinal Stromal Tumors

15. References

16. How Supplied/storage and Handling

17. Patient Counseling Information

* Sections or subsections omitted from the full prescribing information are not listed.

These highlights do not include all the information needed to use Gleevec safely and effectively. See **full prescribing information** for Gleevec.

Gleevec (imatinib mesylate) tablets, for oral use

Initial U.S. Approval: 2001

引自：Highlights of prescribing information: Gleevec—Imatinib mesylate tablet[EB/OL]. [2022-12-20]. https://nctr-crs.fda.gov/fdalabel/ui/spl-summaries/criteria/108276.

Words and Expressions

prescribe	v. 给医嘱，开处方	mesylate	n. 甲磺酸盐

interferon-alpha	α-干扰素	myeloid	*adj.* 脊椎的，骨髓的
myeloproliferative	*adj.* 骨髓增生的	blast crisis	急变期
accelerated phase	加速期	metastatic	*adj.* 转移性的
lymphoblastic	*adj.* 成淋巴细胞的	diuretic	*n.* 利尿剂；*adj.* 利尿的
myelodysplastic	*adj.*骨髓增生异常	chromosome	*n.* 染色体
octanol	*n.* 辛醇	leukemia	*n.* 白血病
crospovidone	交聚维酮	fusion	*n.* 融合
allele deletion	等位基因缺失	ventricular	*adj.* 心室的
silicon dioxide	二氧化硅	malignant	*adj.* 恶性的
heparin	*n.* 肝素	administration	*n.* 给药
contraindication	*n.* 禁忌证	edema	*n.* 浮肿，水肿
magnesium stearate	硬脂酸镁	cytopenia	*n.* 血细胞减少症
refractory	*adj.* 难治的	Warfarin	*n.* 华法林
thyroidectomy	*n.* 甲状腺切除术	rash	*n.* 皮疹
diarrhea	*n.* 痢疾，腹泻		

microcrystalline cellulose	微晶纤维素
gastrointestinal perforation	胃肠道穿孔
gastrointestinal stromal tumors	胃肠道间质瘤
hydroxypropyl methylcellulose	羟丙基甲基纤维素
hypereosinophilic syndrome	高嗜酸性粒细胞综合征
aggressive systemic mastocytosis	系统性肥大细胞增多症
metastatic dermatofibrosarcoma protuberans	隆突性皮肤纤维肉瘤
plateletderived growth factor receptor	血小板衍生生长因子受体

Notes

① Gleevec: 格卫列。"Gleevec"为诺华研发的酪氨酸激酶抑制剂甲磺酸伊马替尼在美国的商品名，其在欧洲、澳大利亚及南美洲被称为"Glivec"。

② Philadelphia chromosome: 费城染色体，简称 Ph。1960 年彼得 · C. 诺埃尔（Peter C. Nowell，1928—2016）和福克斯蔡斯肿瘤研究中心的大卫 · 亨格福德（David Hungerford，1927—1993）发现了慢性粒细胞白血病患者白细胞有一种短小的染色体，称之为费城染色体。

③ c-Kit: 也称为 KIT 或 CD117。c-Kit 是酪氨酸激酶受体蛋白家族的重要成员之一，是一种在造血干细胞以及其他细胞类型表面表达的细胞因子受体，可以通过一系列信号通路参与造血干细胞增殖分化的调控。

④ FIP1L1-PDGFRα fusion kinase: FIP1L1-PDGFRα 融合激酶。PDGFRα 基因重排主要见

于伴嗜酸细胞增多的骨髓增殖性肿瘤，也可见于伴嗜酸细胞增多的急性髓系白血病（无 CBFβ-MYH11 融合基因）或 T 淋巴母细胞淋巴瘤，其中以 FIP1L1-PDGFRα 最为多见。格列卫对伴 FIP1L1-PDGFRα 异常的患者治疗有效。

⑤ CHIC2: cysteine-rich hydrophobic domain 2，富含半胱氨酸的疏水域 2。

⑥ Grade 3/4 hemorrhage: 出血分类 3/4 级。世界卫生组织（WHO）衡量出血严重程度的标准化分级如下：0 级——无出血；1 级——瘀点出血；2 级——轻度失血（临床显著）；3 级——严重失血，需要输血（严重）；4 级——与死亡相关的使人衰弱的失血、视网膜或大脑失血。

⑦ bullous dermatologic reactions: 大疱性皮肤病。是指一组发生在皮肤黏膜以大疱为基本损害的皮肤病，有自身免疫性及遗传性，原发性及继发性之分。

⑧ erythema multiforme: 多形红斑，又称多形渗出性红斑。为急性炎症性皮肤病，有自限性，皮疹多形，有红斑、丘疹、风团、水疱等，特征性皮损为靶形损害即虹膜状皮疹，有不同程度黏膜损害，少数有内脏损害。本病春秋季好发，多发于儿童和青年女性。

⑨ Stevens-Johnson syndrome: 史-约综合征。史-约综合征是一种累及皮肤和黏膜的急性水疱病变。1922 年，首先由 Stevens 和 Johnson 对该病进行了详细的描述。

⑩ TSH: thyroid stimulating hormone，促甲状腺激素。

⑪ CYP: Cytochrome P450 proteins，细胞色素 P450 同工酶。它们是一类主要存在于肝脏、肠道中的单加氧酶，多位于细胞内质网上，催化多种内、外源物质的（包括大多数临床药物）代谢。P450 酶能通过其结构中的血红素中的铁离子传递电子，氧化异源物，增强异源物质的水溶性，使它们更易排出体外。CYP 有多个亚家族，包括 CYP3A4，CYP3A5，CYP2D6，CYP2C9，CYP2C19 等。

Exercises

1. Put the following into English:

慢性粒细胞白血病　　荧光原位杂交　　吞咽困难
体液潴留　　肌肉痉挛

2. Put the following into Chinese:

a. acute lymphoblastic leukemia　　chronic eosinophilic leukemia
hepatic impairment precaution　　cardiogenic shock
renal toxicity　　anticoagulation
film-coated tablets

b. Severe congestive heart failure and left ventricular dysfunction have been reported, particularly in patients with comorbidities and risk factors. Monitor and treat patients with cardiac disease or risk factors for cardiac failure.

c. Hypothyroidism has been reported in thyroidectomy patients undergoing levothyroxine replacement. Closely monitor TSH levels in such patients.

Reading material:

Dying to Survive

Dying to Survive is China's latest blockbuster movie, based on the real-life story of a cancer patient.

In the story, Cheng Yong is the owner of an Indian Miracle Oil Store. He accidentally found that counterfeit leukemia drug Gleevec was only sold at 500 yuan per box in India, compared to over 40, 000 yuan in China. Lured by the greasy profit, Cheng smuggled the unlicensed drug to China and sold to Chinese patients at 2, 000 yuan. He was regarded as a hero by many cancer patients who can't afford the original version of the expensive licensed drug. Cheng became rich and then decided to stop the drug smuggling business. He opened a garment factory. However, a large number of patients became desperate as they were forced to sell their houses or everything they have to pay for authentic Gleevec. That prompted Cheng to change his heart, and renew his smuggling business. This time he even sold the drug only at just 500 yuan to save lives.

Unfortunately, he was arrested and sentenced to jail for five years in the end. A large crowd of patients lined up on the street to show their respect for him as he was being taken away.

Gleevec was developed by Novartis, one of the world's top three pharmaceutical firms. It is considered as one of the best targeted therapy drugs for curing cancer. However, patients need to pay up to US $ 100, 000 for the drug a year without government or insurance subsidy.

In fact, the film was based on the real-life story of Lu Yong, a Chinese textile trader and leukemia patient. He imported less expensive counterfeit drug from India and sold to his fellow Chinese patients. Lu helped thousands of Chinese patients but was arrested and jailed in 2014.

Things are getting better now. Following a medical reform in May 2018, most cancer drugs are now tariff free. More importantly, some of them, including Gleevec, are covered by health insurance schemes that pay up to 80 percent of the costs as subsidy.

China's drug imports soared 16.3 percent to US $ 55.9 billion last year, just behind semiconductor chips and oil. And the imports are set to accelerate after the government scrapped tariff and put anti-cancer drugs into healthcare insurance.

Chapter 5
Translation Techniques for EST

翻译是一种跨语言的交际活动，涉及语言学、社会学、哲学、文学、心理学等领域。翻译不仅是语言之间的转换，还涉及包括历史、地理、风俗、传统、思维方式、价值观念、文学艺术、宗教信仰等文化之间的交流。除了科学、技术、说明、指南等少数文体，大多数文体都带有一定的文化色彩。本书介绍的科技英语翻译技巧，重点为科技文献的翻译技巧。而科技文献的主题是说明新理论、新技术、新方法、新材料、新发现等，所以其语言有很强的客观性、规范性和正式性，用词准确，表达清楚易懂，注重客观叙述，不带感情色彩。科技文献具有专业词汇多、语言简洁明了、逻辑严密等特点，翻译时，灵活性较小，不考虑文化差异，译文要贴近原文。

Unit 1

Introduction of EST

一、科技英语的概念

科技英语(English for Science and Technology,简称 EST)是一种用英语阐述科学技术中的理论、技术、实验和现象等的英语体系,它在词汇、语法和文体诸方面都有自己的特点,从而形成一门专门学科。

科技英语不等于英语语法加科技词汇,不应该把内容结合专业当成科技英语的唯一标志,否则就会误入歧途。科技英语的特点不仅表现在科技内容上,还表现在其特有的语言体系上。

专业英语是科技英语的一部分,以表达科技概念,理论与事实为主要目的。专业英语是结合各自专业的科技英语,具有很强的专业性,表达准确、精练和正式,遵守科技英语的语法体系和翻译方法。

二、科技英语的主要特点

1. 语法特点

科技英语的语法特点包括词类转换多、被动语态多、后置定语多和复杂长句多。

(1) 词类转换多

所谓词类转换,即在翻译时将英文的某种词类译成汉语的另一词类。产生这一现象的根本原因在于英汉两种语言在表达方式上有很大差别,因此在翻译时,应符合各自的表达习惯。例句:

The operation of a machine needs some knowledge of its performance.

操作机器需要懂得机器的一些性能。

原文中的两个名词"operation,knowledge"均译为动词。在科技英语中词类转换的译法十分普遍。

(2) 被动语态多

据统计,被动语态的句子在科技英语中约占 1/3,比非科技英语中被动句出现的频率高一倍左右。这是因为:

首先,科技人员关心的是行为、活动、作用、事实,至于这些行为是谁做的,无关紧要,所以,在这样的句子中就没有必要出现人称。

其次,主语一般位于句首,而被动句是将“行为、活动、作用、事实”等作为主语,因此能立即引起读者对所陈述的事实的注意。

最后,在意思的表达方面,被动句通常比主动句更简洁明了。

试比较例句:

① Mathematics is used in many different fields.

② People use mathematics in many different fields.

不难看出,第一句比第二句更客观、明了。

(3) 后置定语多

后置定语即位于其所修饰的名词之后的定语。汉语常用前置定语或多个简单句来说明某概念或术语,而科技英语则更多地使用后置定语。后置定语的主要形式有介词短语、形容词及形容词短语、副词和分词。例句:

In this factory, the only fuel **available** is coal.

该厂唯一**可用**的燃料是煤。

In addition to aliphatic compounds, there are a number of hydrocarbons **derived from benzene and seemed to have distinctively different chemical properties**.

除了脂肪族化合物以外,还有许多**从苯衍生而来,看来具有明显不同化学性质的**烃。

Besides, isomerization processes may also take place **which in turn leads to other fairly complicated reactions**.

此外,还会发生异构化过程,从而**相继导致其他相当复杂的反应发生**。

(4) 复杂长句多

科技文章要求叙述准确,推理严谨。为了表达清楚,科技英语句子往往较长,有时需认真分析方能明确句子中各成分之间的关系。译成汉语时,必须按照汉语习惯破译成若干简单句,才能条理清楚,避免“欧化句”。

复杂长句是科技英语翻译中的难点之一,它的翻译是建立在词法、句法等基本知识的基础上的。

2. 词汇特点

科技英语中新词、缩略词不断出现,构词手段灵活多样。普遍采用前缀和后缀法构成科技新词。详见第六章。

3. 文章特点

科技英语文章结构严谨,逻辑严密,文体多样,如论文、综述、实验报告、教材、专利、说明书等。

Unit 2
Introduction of Translation

一、翻译的定义

纵观中外翻译发展史,关于翻译的定义可谓琳琅满目,不胜枚举,苏联翻译理论家巴尔胡达罗夫说:翻译是把一种语言的言语产物,在保持内容也就是意义不变的情况下,改变为另一种语言产物的过程。中国现代学者林汉达说:正确的翻译就是尽可能地按照中国语文的习惯,忠实地表达原文中所有的意义。

综合地讲,翻译是用译文语言(target language)把原作语言(source language)忠实地、准确地、完整地再现出来的一种语言活动。从定义可知,翻译时必须确切地理解和掌握原著内容,尽可能保持原文思想,绝不能主观地发挥译者个人的想法和推测,不能有所增删和改变。译者的任务是转换文字而不是改变其意思,要用自己的手段尽可能地将原文的思想清楚准确地表达出来。

二、翻译的标准

翻译的标准是翻译实践的准绳,是评价译文的尺度。自有翻译活动以来,关于翻译标准的讨论就没有停止过。从事佛经翻译的唐朝高僧玄奘提出,翻译"既须求真,又须喻俗",也就是译文既要忠实于原文,又要通俗易懂,这对意义的传达和文字的通达都提出了要求。梁实秋曾经提出"宁错勿顺",鲁迅主张"宁信而不顺"。关于"原文和译文、忠实和通顺"之间的关系一直是翻译标准的一个关键问题。

杨绛在《翻译的技巧》中写道:"翻译是一项苦差事,因为一切得听从主人,不能自作主张。而且一仆二主,同时伺候着两个主人:一是原著,二是译文的读者。"说明译者翻译时既要忠实于原著的思想、情感和风格(要了解原著字句意义以及字句之间与字句之外的含义),又要忠实于读者,用读者的语言传达原著的内容(通俗易懂)。

非科技英语翻译中提倡的是"信(true)、达(smooth)、雅(elegance)"。信即忠实,要求译文和原文是等义的,要忠实原作的内容;达即流畅、通顺,要求译文应当通顺地表达原文的含义,要符合汉语的习惯、表达方式,不能逐字死译。信、达两点是辩证的统一。逐词翻译,以致译文生硬欧化,是应当反对的;单纯追求译文通顺流畅而任意增删原作,也是不正确的。雅即优美,要求译文应当辞藻优美、语言精练,有文学艺术上的价值,使读者获得美的感受。比如:Gone with the wind 译为"乱世佳人";Waterloo Bridge 译为"魂断蓝桥"。

而科技英语翻译提倡“信、达、专业术语正确”，不追求“雅”，而是追求科学性和严谨性。这是科技英语与非科技英语在翻译标准上的区别。进行科技英语翻译时，一定要使用术语，不能说“外行话”。比如：lead structure 译为“先导结构”（不能译为：铅结构）；basic tea solution 译为“碱性茶溶液”（不能译为：基本茶溶液）。

例：The machine works properly.

这台机器运转正常。（误译：机器工作正常。）

解析：work——工作、劳动、做事（指人）；运转、转动、活动（指机械、身体器官等）。

三、翻译的过程

翻译的过程包括对源语的理解和用译出语的表达，大约可以分为以下几个步骤：a. 了解原文的词语、语气、风格、背景等；b. 弄懂词语含义，分析语法和逻辑；c. 寻求贴切的译出语表达；d. 遵照译出语的表达习惯，进行调整、修改、润色。也可以把翻译过程分为三个阶段，即理解、表达和校对。

1. 理解的阶段

通常，翻译时应先通读一遍原文，了解原文的大致内容、所需学科方向以及内容的广度和深度；然后从专业角度细致阅读，针对疑难句子进行语法结构分析，弄清语法关系、句子骨架，扫清陌生或疑难生词，辩明词义，再结合上下文，逐句推敲。

在分析确定词语含义时，需要查阅词典和参考资料，并且通过联系上下文，二者互相辅助，才能选择恰当词义。尤其是遇到专业词汇时，有的可以借助普通词典查找，有的只能在专业词典或其他专业参考资料里查到，还有的无法查到相应的翻译，必须由译者对原文进行分析，自己选择恰当的词汇。

例句：**Oxidation** will make iron and steel rusty.

该句中 oxidation 具有“氧化现象、氧化作用、氧化过程”等含义。本句在讲氧化（　　）使钢铁生锈问题，根据常识可知，氧化作用可以使钢铁生锈，钢铁生锈是氧化现象故应该选择其词义为“氧化作用”，本句译为：氧化作用会使钢铁生锈。

2. 表达的阶段

表达是选择恰当的汉语，把已经理解了的原作内容叙述出来。表达是翻译过程的关键，是最终决定译文优劣的关键。而表达的好坏取决于理解原文的确切程度和对译出语的掌握程度。

理解和表达不能截然分开，理解原文的过程也是形成表达的过程。但是，由于英语和汉语在文化上的差异、在表达习惯上的不同以及相应的词语所具有的不同隐含意义，在表达的时候，除了要贴近源语含义，还要遵照译出语的表达习惯。如果译文仅仅是意思对，但不能用通顺流畅的汉语表达，仍不是一篇好译文。

例句：The homologs of benzene are those containing an alkyl group or alkyl groups in place of one or more hydrogen atoms.

这句话容易理解,但是不好表达。通常会被翻译为:苯的同系物就是那些被一个或多个烷基取代一个或多个氢原子所形成的产物。虽然该译文意思差不多,但令人感到啰唆、费解,仍不是好译文。

正确译文:苯的同系物是那些苯环上含有单烷基(取代一个氢)或多烷基(取代多个氢)的物质。

3. 校对的阶段

校对工作对提高译文质量起着很大的作用。通过校对可以发现在理解阶段或表达阶段不够确切或不完整的地方,可以检查译文是否能准确无误地转述原作内容、译文的语言表达是否规范、是否符合汉语习惯。

通常校对修改需要在誊写工整的译文上进行。一校后誊写清楚,然后再进行修改校对,如此重复,直到满意为止。因此,翻译不是一种机械劳动,而是复杂的创造性的脑力劳动,只有熟悉汉英两种语言并掌握了翻译方法之后,才能胜任英汉语翻译工作。

Unit 3
Translation Techniques

英汉两种语言属于两种完全不同的语系,无论在词汇,还是在语法结构上都存在着巨大差异,因此,英汉两种语言的句子在措辞、语序、结构等方面各有特点,如英语多后置定语,而汉语多前置定语;英语多长句(句子中心在句首),而汉语多短句(句子中心在句尾);英语多非人称做主语,而汉语多以人称做主语;英语中状语多放于谓语动词之后,而汉语则习惯放动词之前;英语多被动语态,汉语多主动语态。因此,对英文文献的翻译并非简单对应的词汇罗列。

对于科技英语而言,译者应更注重逻辑思维,讲究语言规范,表达妥帖,译文流畅,措辞严谨,不拖泥带水。另外,必须掌握相当的专业知识,恰当地运用专业词汇及专业术语。由于英文科技文献具有思维严密、措辞严谨、句子结构复杂等特点,因此,掌握一定的翻译技巧是非常必要的。

任何翻译技巧都不是放之四海皆准的法则。鲁迅所倡导的“译事双无”中的“无”就是“译无定法”。多数情况需要译者综合、灵活地运用各种翻译技巧。

一、词义选择法

根据英汉两种语言的一词多类、一词多义现象,在翻译过程中,需要根据上下文、词在句中的词类、词的搭配个性来选择恰当的词义。

1. 根据上下文来选择词义

同一个词,词类相同时,在不同的语言环境中往往具有不同的含义,必须根据上下文来判断和确定该词在特定语言环境中应具有的词义。

以 operate 为例:

The ability to operate at high frequency is important characteristics of microwave device.

高频**运转**能力是微波器件的一个重要特性。

The relay is operated by a current of several milliamperes.

继电器由数毫安电流**启动**。

There is no doubt that we will be able, in the next generation, to build jet transports operating at speeds of one or two thousand miles an hour.

毫无疑问,我们在未来的二三十年内,将能建造每小时**运行**一二千英里的喷气式运输机。(句中 the next generation 采用引申法,译为“未来的二三十年”)

2. 根据搭配关系来选择词义

汉英两种语言在长期使用过程中形成了各自的固定词组和搭配用法，翻译时必须注意两者的不同，不能生搬硬套。

heavy	heavy fate	苦命	heavy taxes	苛税
	heavy task	繁重的任务	heavy machine	重型机械
	heavy traffic	拥挤的交通	heavy sea	波涛汹涌的大海
	heavy current	强电流	heavy makeup	浓妆
build	build a fire	生火	build a ship	造船
	build a dam	筑坝	build a house	盖房子
青	青山	green hill	青天	blue sky
	青布	black cloth	青草	green grass

Note: 2005 年习近平总书记首次提出“绿水青山就是金山银山”的重要论述。强调人与环境的和谐。

开	to set sail	开船出海	to start a machine	开机器
	to open a door	开门	to drive a car	开车
大	大考	final exam	大手术	major operation
	大丈夫	true man	大风	strong wind
	大浪	high wave	大清早	early morning
	大雨	heavy rain	大白天	in broad daylight
红	红茶	black tea	红糖	brown sugar
	红眼病	pink eyes/be green with jealousy		
浓	浓茶	strong tea	浓墨	dark/thick ink
	浓烟	dense smoke	浓眉大眼	bushy brows and big eyes
	浓厚的兴趣	great interest		

3. 根据词类选择词义

一词多类多义时，首先确定该词在句中属于哪一词类，然后根据词类选择一个确切的词义。如 round:

The night watchman makes his rounds every hour.

夜间值班人员每小时**巡视**一周。(*n.* 巡视)

The earth rounds the sun.

地球**围绕**太阳**运行**。(*vt.* 绕行)

It is neither round nor square.

它既不**圆**也不方。(*adj.* 圆的)

The wheel turns round.

轮子**循环**转动。(*adv.* 循环地)

When direct current flows through a coil, a magnetic field will be built up round the coil.

当直流电通过线圈时，该线圈**周围**就形成一个磁场。(*prep.* 在……周围)

二、引申法

英语词汇的内涵和外延跟汉语有很大的差别。在英汉互译时，照字面意思翻译，必然晦涩难懂，容易引起诸多误解。因此，必须根据词语所处的语言环境，从意义上、语气上、逻辑关系及搭配关系上引申出能表达词的内在含义的新词义。

1. 通过表象揭示内涵

有些词按字典上查到的意义进行翻译，译文往往不通顺，不能表达原作的确切含义，甚至令人费解。此时，译者应根据该词本意，结合上下文，将原文的词语加以引申，选择更恰当的词义来揭示内涵。例如：

To implant a pace-maker anywhere in the body is to ask the tissues to carry an extra and **unnatural** weight.

在体内某处安放心脏起搏器，就是要组织承担一份额外**非身体的**重量。

The two pairs of electrons of oxygen may be shared with two **separate** carbons forming only single bond.

氧的两对电子可以与两个**不直接相连的**碳共用而形成单键。

The continuous process can be conducted at any **prevailing pressure** without **release to atmospheric pressure**.

连续过程能在任何**常用的**压力下进行，而不会**暴露在大气中**。

2. 将概括的词义做具体化引申

根据语言习惯，将原文中比较概括、抽象的词引申为较具体的词。例如：

There are many **things** to be considered in determining cutting speed.

在确定切削速度时，有许多**因素**应当考虑。(把抽象名词“事情”引申为“因素”)

The major problem in fabrication is the control of contamination and **foreign materials**.

加工中存在的主要问题是如何控制污染和**杂质**。

3. 将具体的词义做抽象化的引申

把原文中词义较具体的词引申为词义较抽象的词，或把词义较形象的词引申为词义较一般的词。例如：

Steel and cast iron also differ in **carbon**.

钢和(铸)铁的**含碳量**也不相同。

There are three steps which must be taken before we **graduate from** the integrated circuit technology.

我们要完全**掌握**集成电路工艺学，还需经过三个阶段。

三、增词法

由于文化背景的差异,中英文语言表达方式不同,在翻译过程中为了使译文通顺达意,而在译文中增补某些必要的词,即为增词法。所增补的词必须是意义上、句法上以及搭配上所需要的,不可无中生有、随意增补。

1. 由于意义上的需要而增词

(1) 根据意义上的需要可在英语的名词、动名词前增加动词和形容词。例如:

Some **process**, such as oil refining, may be completely automated from start to finish.

有些**生产过程**,如提炼石油,可以全部自动化。

(在名词 process 前增加动词“生产”)

Speed and reliability are the chief advantage of the electronic computer.

速度**快**、可靠性**高**是电子计算机的主要优点。

(在名词 speed 和 reliability 后增加形容词“快、高”)

(2) 当含有动作意义的抽象名词表示具体概念时,常可通过增词法使译文具体化。这类词通常有:作用、现象、效应、方法、过程、工作、装置、情况、设计、变化、加工、结构、结果等。例如:

Oxidation will make iron and steel rusty.

氧化作用会使钢铁生锈。

From the **evaporation** of water, people know that liquids can turn into gases under certain condition.

从水的**蒸发现象**,人们得知液体在一定条件下能转变成气体。

(3) 当英语的某些形容词单独译出意思不明确时,可在其前增加名词使译文意思明确。例如:

A new kind of aircraft—**small**, **cheap**, pilotless, is attracting increasing attention.

一种新型的飞机正越来越引起人们的注意——这种飞机**体积**小,**造价**低,而且无人驾驶。

According to Newton's Third Law of Motion, action and reaction are **equal** and **opposite**.

根据牛顿运动第三定律,作用力和反作用力是**大小**相等,**方向**相反的。

(4) 英语中常因惯用法和上下文关系,省去了不影响理解全句意义的词语。但译成汉语时必须增译这些省去的词语,否则译文不清楚。例如:

We will go to the Wuhan Bridge by a **Liberation**.

我们将乘解放牌**汽车**去武汉大桥。

But neutralization is **cumbersome**.

但是,中和是个**麻烦的问题**。

2. 由于句法上的需要而增词

（1）在英语的并列句中，后一个分句常常省略某些与前一个分句相同的词。英译汉时，为使语义清楚，需增补所省略的词。例如：

This year **the production** of iron in their plant **has increased** by 15 per cent, and of steel by 13 per cent.（后一个分句省略了 the production 和 has increased）

今年，他们厂的铁产量增长了15%，**钢产量增长**了13%。

The letter *I* **stands for** the current in ampere, *E* the electromotive force in volts, and *R* the resistance in ohms.（后面分句 *E* 和 *R* 后省略了 stands for）

字母 *I* 代表电流的安培数，*E* **代表**电动势的伏特数，*R* **代表**电阻的欧姆数。

（2）when, while, once, if, unless 等从属连词常直接跟分词、形容词或介词短语等一起构成状语。这类词语可视为"省略性的状语从句"。它的逻辑主语与主句的主语相同，翻译时，要增补从句的主语，主句的主语有时译出，有时省略。例如：

When pure, water is a colourless liquid.

水纯时，水是无色的液体。

Turning around, she saw Tom in tears.

她回过头，看到汤姆在流泪。

（3）英语被动句中，如谓语是表示"知道、了解、看见、认为、发现、考虑"等意义的动词时，通常可增加"大家、人们、我们、有人"等，译为主动句。例如：

All matter is **known** to possess weight.

大家知道，物体都有重量。

Many elements in nature are **found** to be mixtures of different isotopes.

人们发现，自然界有许多元素都是由各种不同的同位素混合组成的。

3. 由于搭配上的需要而增词

由于搭配上的需要而增词的例子很多，必须根据不同的搭配而译出确切句义。

（1）形容词最高级作定语的句子译成汉语时，由于搭配上的需要，常增补带有"让步"意义的词。例如：

The **strongest** spring leads to fatigue failure caused by excessively high stress.

即使强度最大的弹簧，由于应力过大，**也**会导致疲劳损坏.

The **best** lubricant can not maintain oil films between the surfaces of engineering gears.

即使最好的润滑油，在啮合处**也**不能保住油膜。

（2）在 could...with 结构中，须增补带有"条件"意义的词。例如：

The life **could** be significantly improved **with** the good use of coating.

如果正确地使用涂料，使用寿命还可显著延长。

The atomic reactor **could** not run **with** too many neutrons.

倘若中子太多，原子反应堆就无法控制。

（3）介词 with 短语可增补带有"让步"意义的短语。例如：

With all his achievements, he remains modest and prudent.

他**虽**有很多成就,但还是谦虚、谨慎。

With your eyes shut, you can recognize hundreds of things by their sound or touch.

虽然闭着眼睛,但还能凭其声音或触感辨认几百种东西。

四、减词法

在英汉互译中,由于英汉两种语言在词义范围和使用习惯上、语法体系和结构上以及搭配上的差异,为使译文准确、流畅,在互译中将不符合两种语言习惯的词语省去,称为减词法。

1. 由于词义范围和使用习惯不同而减词

(1) 无论英译汉,还是汉译英,有时由于词义范围和使用习惯的不同,必须将英语中的 both, this, these, all 以及汉语中的"这些、所有、双方、等"之类的词语在互译中省去不译。例如:

Both John and Mary left for Beijing yesterday.

约翰和玛丽昨天去北京了。(省译了 both)

All of my friends like riding.

我的朋友都喜欢骑马。(省译了 all)

(2) 英语介词 of 前表示度量等意义的名词及同义名词,可省略不译。例如:

Going through **the process** of heat treatment, metals become much stronger and more durable.

金属经过**热处理**后,强度更大,更加耐用了。

When current flows in a conductor, the electrical energy converted into heat is directly proportional to **the length** of time the current flows.

电流通过导体时,转换成热能的电能与通电的时间成正比。

(3) 英语中有些动词如联系动词 be、get、become 等,还有实义动词 cover、carry 等在翻译时可省略。例如:

When the pressure **gets** low, the boiling point **becomes** low.

气压低,沸点就低。

This truck **carries** a maximum load of five tons.

这辆卡车的最大载重量为五吨。

2. 由于语法体系和结构不同而减词

(1) 由于汉语中没有冠词,在英译汉时,除某些情况外,基本上将冠词省略。例如:

Hydrogen is **the** lightest element.

氢是最轻的元素。(省译形容词最高级前的定冠词)

(2) 英语中有些代词在译成汉语时必须省略,以使译文简洁,句义明白无误。例如:

We cannot see sound wave as they travel through air.

声波在空气中传播时，是看不见的。

3. 由于搭配不同而减词

英文中有些词、词组，由于同汉语在词的搭配关系上不同，可以不译出，如译出，意思反而令人费解。例如：

The temperature in the sun is very high; **that is the reason why** nothing can exist in solid state.

太阳中的温度很高，**因此**没有任何东西能以固体存在。

The helium atom has two electrons outside its nucleus, **which accounts for the fact that** its chemical properties differ from those of hydrogen.

氦原子的原子核外有两个电子，**所以**它的化学特性不同于氢原子。

五、转换法

由于汉英两种语言在结构和表达方式上千差万别，在翻译时就不能逐字死译，否则无法表达原文的意思。做好翻译的关键是找到两种语言表达方式的差异，再根据各自语言规范进行翻译。所以，英译汉时，译者应大胆摆脱原文表层结构的束缚，根据汉语习惯，正确表达原文，这就要求译者必须进行必要的转换或变通，才能做到既忠实于原文又符合译文规范。

翻译实质上是两种语言的转换，没有转换，就没有翻译。转换是一切翻译的理论与技巧的基石。转换译法是一种可取的、极为常用的翻译方法。

转换译法包括词类的转换、句子成分的转换、词序的转换、主被动语态的转换、从句间的转换等。本书重点阐述词类转换法。词类的转换就是翻译时改变原文中某些词的词性，以符合汉语的表达习惯。下面主要介绍名词的转换、动词的转换、形容词的转换、副词的转换、介词的转换等。

1. 名词的转换

（1）英语中由动词转换过来的名词或具有动作意义的名词，可以译成汉语的动词，即英语名词转换为汉语动词。例如：

Total **determination** of molecular structure is possible by means of X-ray diffraction.

用 X 射线衍射的方法可以全面地**确定**分子结构。

Such operations require the **transfer** of a substance from the gas stream to the liquid.

这样的一些操作要求物质由气流**传递**到液体中去。

（2）英语中有些加后缀的名词，有时在句中并不指其身份和职业，而是含有较强的动作意味，往往可译成汉语动词。例如：

I am afraid I can't teach you swimming; I think my little brother is a better **teacher** than I.

恐怕我教不了你游泳，我想我的小弟弟比我**教**得好。

He is no **smoker**, but his father is a **chain-smoker**.

他不**抽烟**,但他的父亲**烟抽得**很重。

(3) 一些情况下,英语名词可以转译为汉语形容词。例如:

He is **a stranger** to the operation of the electronic computer.

他对电子计算机的操作是陌生的。

The maiden voyage of the newly-built steamship was a **success**.

那艘新轮船的处女航是**成功的**。

2. 动词的转换

(1) 英语中,有些名词派生的动词以及由名词转用的动词,在汉语中往往不易找到相应的动词。翻译时,可将其转译成名词。例如:

Mantle cell lymphoma is **characterized** by inactivation of the ATM gene.

袍状细胞淋巴瘤的**特点**是钝化 ATM 基因。

Protein synthesis in eukaryotic cells **differs from** that...

真核生物细胞中蛋白质合成的**区别**在于……

It is known that neutrons **act** differently from protons.

大家知道,中子的**作用**不同于质子。

(2) 由于汉语表达习惯的限制,可以把英语动词转译为汉语副词。例如:

The methyl group on the benzene ring greatly **facilitates** the nitration of toluene.

苯环上的甲基使甲苯非常**易于**硝化。

Since the 1940s, many other fermentations have been **commercialized**.

20 世纪 40 年代以来,许多其他发酵产品都已实现**商业化**生产。

3. 形容词的转换

(1) 英语中有一些表示人和事物特征的形容词作表语用时,可在其后面加"体、性、度"这类词,译成汉语的名词。例如:

Generally speaking, methane series are rather **inert**.

总的来说,甲烷系的烃**惰性**很强。

Zirconium is almost as **strong** as steel, but lighter.

锆的**强度**几乎与钢的相等,但它比钢轻。

(2) 英语中表示知觉、情感、欲望等心理状态的形容词在句中作表语使用时,译成汉语的动词。例如:

They are quite **content** with the data obtained from the experiment.

他们对实验中获得的数据非常**满意**。

Also **present** in solids are numbers of free electrons.

固体中也**存在着**大量的自由电子。

4. 副词的转换

英语中副词可以转换成汉语的动词、名词、形容词等。例如:

The chemical experiment is **over**.

化学实验**结束了**。(英语副词转换为汉语动词)

Oxygen is very active **chemically**.

氧的**化学性能**很活泼。(英语副词转换为汉语名词)

The image must be **dimensionally** correct.

图的**尺寸**必须正确。(英语副词转换为汉语名词)

This film is **uniformly** thin.

该膜薄而**均匀**。(英语副词转换为汉语形容词)

5. 介词的转换

有些具有动作意味的介词,如 for、by、in、past、with、over、into、around、across、toward、through、throughout 等,可转译为汉语动词。例如:

In this process, the solution is pumped **into** a tank.

在这个操作中,溶液用泵**打入**罐中。

Since most drugs are absorbed by passive diffusion **through** the lipoid barrier, the lipid/water partition coefficient and the pK_2 of the drugs are of prime importance to...

由于大部分药物是通过被动扩散**穿过**类脂屏障的;药物的油脂分配系数与 pK_2 值对……是非常重要的。

六、被动语态的译法

英语和汉语都有被动语态,但在语态使用上存在一定差异。汉语中被动语态使用得很少,而英语中被动语态的使用非常广泛,尤其在科技英语中的使用频率特别高,据初步统计,被动语态在科技英语中约占 1/3,远高于普通英语中被动句的使用频率。这是因为:

i. 与主动语态相比,被动语态会突出所要说明的事物,能减少主观色彩。因为科技人员关心的是行为活动、作用、事实,至于这些行动是谁做的,无关紧要,因此在这样的句子中就没有必要出现人称。

ii. 被动句用行为、活动、作用、事实等作主语,在句中是第一个出现的词,因此能立即引起读者的注意。

iii. 通常被动句比主动句更简洁明了。

翻译被动语态时,可根据具体情况翻译成被动句或主动句。

1. 仍译成汉语的被动句

汉语中被动关系的表达方式很丰富,可在谓语前加译“被、于、供、由、给、让、遭、受、得到、为……所、予以、加以”等助词。这些助词在汉语中都表示被动的意思。有时也可不加任何词而直接译出。翻译时要根据汉语的表达习惯选用恰当的表达方式。例如:

Purified hydrogen is passed over a liquid metal halide.

让纯化氢气通过液态金属卤化物。

Since methane is its first member, it is also known as the methane succession/series.
因为甲烷是该系列的第一个成员，因此该系列也被称为甲烷系。

2. 译成汉语主动句

（1）译成无主语句：英语中有些被动句不需要或无法说出行为的主体，翻译时往往译成无主语句。一般是将英语句中的主语译成汉语中的宾语，即先译谓语，后译主语（主谓变序）。例如：

When the solution in the tank has reached the desired temperature, **it** is **discharged**.
当罐内溶液达到所要求的温度时，就**卸料**。

If the product is a new compound, the **structure** must **be proved** independently.
如果产物是一个新的化合物，则必须单独**证明其结构**。

Measures have **been taken** to prevent the epidemic from spreading quickly.
已经**采取了一些措施**来防止这种流行病迅速蔓延。

（2）将英语句中的一个适当成分译作汉语句中的主语。例如：

The oxides of nitrogen are absorbed in **water** to give nitric acid.
水吸收氮氧化物生成硝酸。

The following five groups of antibiotics represent a yearly worldwide value of $ 17 billion.
下面五类抗生素的世界年产值是170亿美元。

（3）在原主语前加译“把、将、给、使”等词。例如：

Borax was added to beeswax to enhance its emulsifying power.
将硼砂加到蜂蜡中以增加蜂蜡的乳化力。

Absorption process is therefore conveniently divided into two groups: physical process and chemical process.
可以**将**吸收过程简单地分为两类：物理过程和化学过程。

（4）被动语态着重于对事物的状态、过程或性质等进行描写，其作用与系表结构类似，可翻译成汉语中的判断句：“是……的；对……进行”等。例如：

An aldehyde is prepared from the dehydrogenation of an alcohol and hence the name.
醛**是**从醇脱氢而制备**的**，并因此而得名。

The slurry is filtered to recover the electrolyte solution.
对泥浆**进行**过滤可回收电解液。

3. 被动意义译成主动意思

如果被动语态体现的是主语的情感或感受，或并非强调被动动作，或译成汉语时没有表示被动关系的词也通顺，则可以保持主语不变，省略体现被动关系的词。例如：

Batch operations are frequently **found** in experimental and pilot-plant operations.
间歇操作**常见于**实验室操作及中试操作。

In the reaction, both the acid and the base are **neutralized** forming water and salt.
反应中酸与碱**彼此中和**而形成水和盐。

七、长句的译法

在表达比较复杂的概念时，英汉两种语言差别较大。英语习惯利用各种修饰语，构成较长的简单句，再利用适当的连词将简单句构成更长的并列句或复合句。因此科技英语中常使用长句子。而汉语则倾向于尽量使用若干短句来说明某个概念，不常使用较长的句子。英译汉时，要特别注意英语和汉语之间的差异，将英语长句破开成几个汉语短句，然后再根据逻辑次序和意思轻重，重新安排句子结构。

英语长句分析是科技英语长句翻译的关键。科技英语中的难句主要表现在句子长、结构复杂。其实，不管句子有多么长，多么复杂，都有一定的规律可循。这些复杂的句子都是由主语、谓语、宾语等基本成分组成，分析长句要把握以下几个要素：(1) 分析全句的主干，找出全句的主语、谓语和宾语，从整体上把握句子结构；(2) 找出句中所有的谓语结构、非谓语动词；(3) 找出句中所有的长句引导词，并分析从句的功能；(4) 分析词、短语和从句之间的关系。例如，定语从句所修饰的先行词是哪一个？句中是否有固定词组或固定搭配？翻译英语长句的过程是一个综合运用语言和综合运用各种翻译技巧的过程。

1. 顺译法

当英语长句内容的叙述层次与汉语基本一致时，可以按照英语原文的顺序翻译成汉语。例如：

Gas absorption is an operation in which a gas mixture is contacted with a liquid for the purposes of preferentially dissolving one or more components of the gas and to provide a solution of these in the liquid.

分析：① 本句为复合句；② 定语从句译成并列分句；③ 被动语态 is contacted 译为“让气体混合物与液体接触……”。

译文：气体吸收是这样一种操作——让气体混合物与液体接触，以使气体中的一种或多种组分优先溶解在液体中，并提供由这些组分和液体所形成的溶液。

The continuous process although requiring more carefully designed equipment than the batch process, can ordinarily be handled in less space, fits in with other continuous steps more smoothly, and can be conducted at any prevailing pressure without release to atmospheric pressure.

分析：① 本句为并列句；② 句中 less 为形容词，采用词类转换译法，转译为动词“节省”；③ 最后一个分句采用引申法，译为“并能在任何常用的压力下进行，而不会暴露在大气中”。

译文：虽然连续过程比间歇过程需要更为周密设计的设备，但连续过程通常能节约操作空间，较顺利地适应其他连续操作步骤，并能在任何压力下进行，而不会暴露在大气中。

2. 倒译法(变序译法)

有时英语长句的表达顺序与汉语表达习惯不同，甚至大相径庭。比如，英语中最重要的信息往往出现在句首,然后再交代条件、原因等,而汉语通常先说明条件、背景等,然后再出现最重要或最新信息,或者按照逻辑关系和时间先后安排语序。因此,英译汉过程中,应根据汉语表达习惯,改变原文语序,进行翻译。这种译法为倒译法,又称变序译法。

倒译法通常在下列情况下使用:① 主句后面带有很长的状语(特别是原因状语或方式状语)或状语从句(特别是原因、条件、让步状语从句);② 主句后面有很长的定语或定语从句,或宾语从句,而后者按汉语习惯常常译在主句之前时。例如:

This is why the hot water system in a furnace will operate without the use of a water pump, if the pipes are arranged so that the hottest water rises while the coldest water runs down again to the furnace.

分析:先译 if 引导的条件状语从句。

译文:如果把管子装成这个样子,使最热的水上升,而最冷的水往下回流到锅炉里,那么锅炉中的热水系统不用水泵就能循环。

We learn that sodium or any of its compounds produces a spectrum having a bright yellow double line by noticing that there is no such line in the spectrum of light when sodium is not present, but that if the smallest quantity of sodium be thrown into the flame or other sources of light, the bright yellow line instantly appears.

分析:从"by noticing that"一直到结尾都是主句的方式状语。该状语既长,又较复杂,状语中又有从句,应先译较长的方式状语,再译主句。

译文:我们注意到,如果把非常少量的钠投入火焰或其他光源中,立即会出现一条亮黄色的双线,而当钠不存在时,光谱中就没有这样的双线。由此我们知道钠或钠的任何化合物所产生的光谱都带有一条亮黄色的双线。

3. 分译法

当英语长句中并列成分较多,或主句与从句、短语与修饰词之间的关系不十分密切,各具相对独立性时,可以按照汉语多用短句的习惯,把长句拆分成多个短句译出,有时还需要适当增加词语或调整语序。这种翻译方法即为分译法。

分译法通常是将原文的某一短语或从句先行单独译出,并借助适当的总括性词语或其他语法手段将前后句联系到一起;或将几个并列成分先概括地合译在前面,而后分别加以叙述;或将原文中不好处理的成分拆开,译成相应的句子或另一独立句子。例如:

Our object is to draw attention to those areas responsible for the quantum jump in sophistication, improved cosmetic attributes and safety of today's products.

分析:① 添加总括性词语"下列方面"并先行译出,而后具体叙述;② to draw attention to 译为"提请人们注意"(不能译作:我们注意)。

译文:我们的目的是提请人们注意那些使下列方面发生跃变的领域,即当代化妆品所采用的先进技术方面、质量改进方面以及安全性方面。

Another development, by itself not strictly scientific in nature, is the growing awareness that the energy resources at present available to human technology are limited, and that photosynthesis is the only large-scale process on earth by which a virtually inexhaustible source of energy, i.e., the radiation energy of the sun, is collected and converted into a form of energy that is not only used by plants, but by all forms of life, including man.

分析:先加译总括性词语"下面这个事实",再分别叙述。

译文:另一发展——就其本性而言,本身不是严格科学的——是人们正日益认识到下面这个事实,即:人类技术领域目前可以利用的能源是有限的,在地球上只有光合作用是能将那实际上是取之不尽用之不竭的能源——太阳的辐射能聚集起来,转换为不仅植物,而且包括人在内的所有生命都可以利用的能量形式的大规模过程。

事实上,在翻译英语长句时,并不只是单纯地使用哪种单独的译法,而是需要综合使用顺译法、倒译法和分译法。

Chapter 6
Specialized English Word-Building Method

词是构成句子的要素,对词义理解得好坏直接关系到翻译的质量。对于制药专业学生而言,需要掌握的专业词汇涉及面比较广,包括化学、生物学、药学、工程学等多学科专业词汇。但通常不需要死记硬背所有这些专业词汇,仅需要熟练掌握一些常用的专业词汇以及在构词上有很大作用的词根与前后缀即可。而这些词根、前后缀构成的单词往往在结构上具有一定的规律,这些规律即为构词法(词的构成方法)。

专业英语构词法对专业英语词汇的丰富和发展起到了重要作用,它是专业词汇学习的重要方法之一,它研究词汇本身的内部结构,从词汇的构成本源出发,将词汇的构词分为多种方法,除了非科技英语中常用的三种构词法——转化法、派生法及合成法外,还普遍采用压缩法、混成法、符号法和字母象形法。其中,最重要的就是派生法,它是大多数专业词汇的主要构词法。

Unit 1
Conversion for Part of Speech

词类转化法(conversion for part of speech)是由一个词类直接转化为另一个词类,而不改变其词形的构词方法,即某一个词早期迁移定型为一个词义和词类,后来通过词义扩大或比喻等方法把它活用为其他词义和词类的方法,一般在转化义和原义之间有明显的联系。英语中,词类转换法是一个特别多产的构词法。

词类转化法多用于简单音节的词汇,常见为名词转换为动词、形容词转换为动词、形容词转换为名词、动词转换为名词、名词转换为形容词等。如:

charge(*n.* 电荷)→charge(*v.* 充电)

yield(*n.* 产率)→yield(*v.* 生成)

slow(*adj.* 慢的)→slow(*v.* 减慢)

back(*adv.* 在后、向后)→back(*v.* 使后退、倒车)

square(*n.* 正方形)→square(*adj.* 正方形的)

swim(*v.* 游泳)→swim(*n.* 游泳)

red(*adj.* 红色的)→red(*n.* 红色)

Unit 2
Derivation

派生法(derivation)是指以一个词根为词干,借助某些前缀或者后缀,从而构造出一个新词汇的方法。英语中绝大多数单词是由添加词缀而来的派生词,其基本部分为词根,表达了该词的基本含义。由该词根可派生出大量的单词,因而掌握了词根的基本含义以及词缀的意义,便可熟记大量的单词。词根、前缀、后缀称为扩大词汇的三把钥匙,由此可见,派生法在新词构建中具有重要作用。

派生法也是科技英语的重要构词方法。科技英语中许多比较长的词都是通过词根加词缀构成的,翻译时,可以通过构词规律,采用先解析、后合成的方法,得出一个完整的词义。如 tetracycline=tetra(四)+ cycl(环)+ine(表示"药物"),译为四环素。据估计,通过一个前缀可帮助人们认识 50 个左右的英语单词。

例如,-ee 是名词后缀,当把其加在某些动词之后,就可以构成一个带有被动意义的新名词:

trust	trustee	受信托者	employ	employee	被雇佣者
expel	expellee	被驱逐者	interview	interviewee	被接见者

一名科技工作者至少要知道近 50 个前缀和 30 个后缀。利用这种方法,可以顺利解决许多词典上查不到的专业词汇,对扩大科技词汇量、增强自由阅读能力、提高翻译质量和加快翻译速度都是大有裨益的。

一、常见的前缀

Prefix (meaning or function)	Examples		
a-(an-) 不,无,非	atypical 不典型的	anarchy 无政府状态	anhydrous 无水的
amino- 氨基	amino acid 氨基酸	2-amino ethanol 2-氨基乙醇	*p*-aminophenol 对氨基苯酚
amphi-两, 双, 指萘环的 2, 6 位	amphiphilic 两亲的, 两性分子的	amphion 两性离子	amphi-naphthoquinone 2, 6-萘醌
Note: quinone (苯)醌,其中 quin 音译为醌;-one 表示酮,含有羰基的化合物			

continued

Prefix (meaning or function)	Examples			
anti- 反,抗,阻,排斥	anti-aging agent 防老剂	antibody 抗体	antidote 解毒剂,解药	
	antineoplastic agents 抗肿瘤药	antimetabolic 抗代谢的	antipyretic 解热的	
Note: agonist 兴奋剂，激动剂；antagonist 拮抗剂；partial agonist 部分激动剂				
acyl- 酰基	acylglycerol 酰基甘油	acyl chloride 酰基氯		
auto- 自己,自动	automatic 自动化的	autobiography 自传	autoantibiosis 自体(/动)抗菌作用	
baro- 重,压	barometer 气压计	barite 重晶石		
bi-两，二元的，酸式的，重	bichromate 重铬酸盐	bicarbonate 重碳酸盐,碳酸氢盐	bicyclic 二环的	
centi- 百分之一,厘	centimeter 厘米	centipoise 厘泊		
cis- 顺式	cis-1，3-butadiene 顺-1，3-丁二烯	cis-form 顺式		
contra-;counter- 反对,逆,抗,相反	contra-rotating 反向旋转	counterclockwise 逆时针方向的		
	counter-electrode 反电极	counteract 抵抗，中和		
de- 除去,离去,非	dehydrate 脱水	deforest 砍伐森林	deionized water 去离子水	denature 变性(蛋白质、核酸等)
Note: deoxyribonucleic acid 脱氧核糖核酸(DNA)(oxy 氧;ribo 核糖;nucleic acid 核酸)				
deci- 十分之一	decimeter 分米	decimal 十进制的		
di- 二,双,二倍	dioxide 二氧化物	butenediol 丁烯二醇		
di-; dif-; dis- 不,无,拆开	digest 消化	different 不同的	disappear 消失	

continued

Prefix (meaning or function)	Examples				
equ- 等,均,平,同	equal 相等的	equation 方程式,等式,反应式,公式	equilibrium 平衡		
glyco-/glyc- 糖,甘,甜	glycogen 糖原	glycoprotein 糖蛋白	glycine 甘氨酸		
hemi-; semi- 半,二分之一	hemiacetal 半缩醛	hemiterpene 半帖	hemicolloid 半胶体		
hept(a)- 七	heptane 庚烷	heptatomic 七原子的,七元的			
hetero- 异,杂	heterocycle 杂环	heteroatom 杂原子	heterogeneous 多相的,非均相的		
hex (a)- 六	hexene 己烯	hexavalence 六价			
homo- 相同,匀	homogeneous 同族的, 均相的	homopolar 同极的	homophase 同相		
hyper- 超过,在上,过多,高	hypertension 高血压	hyper cholesterol 高胆固醇	hyperglycemia 高血糖		
hypo- 在下,次,低	hypotension 低血压	hypochlorite 次氯酸盐			
im-;in-;ir- 不,无	impermeable 不可渗透的	inactive 不活泼的	inorganic 无机的	irreversible 不可逆的	
infra- 下,低,次,外	infrared 红外线	infrasonic 低于声频的			
inter- 中间, 相互	intermolecule 分子间	international 国际的	interchange 互换,变换		
intra- 内,在内,内部	intramolecule 分子内	intrastate 本州内的			
Note: molecule *n.* 分子; molecular *adj.* 分子的; intramolecular 分子间的					
iso-相同,相等, 异(有机物命名)	isotope 同位素	isobar 等压线	isoelectronic 等电子的	isooctane 异辛烷	isopolyacid 同多酸
Note: bioisostere principle 生物电子等排原理					

continued

Prefix (meaning or function)	Examples			
kilo- 千	kilogramme 千克,公斤	kilometer 千米	kilovolt 千伏特	
macro- 宏(观),大(量)	macromolecule 大分子	macroworld 宏观世界		
meta-(m-) 变化,间位,偏(无机酸)	metamorphic 变形的	meta-compound 间位化合物	meta-acid 偏酸	
micro- 微,小,百万分之一	microscope 显微镜	microgramme 微克		
milli- 千分之一,毫	milligramme 毫克	milliliter 毫升		
mono- 单,一	monoester 单酯	monohydrate 一水合物	monoacid 一元酸	
multi- 多	multivalence 多价	multiphase 多相的	multicomponent 多组分的	multilayer 多层,多分子层
myco-/mycete- 霉菌,分枝,真菌	mycology 真菌学	mycobacterium 分枝杆菌	streptomycete 链霉菌	
non(a)- 不,无,非	nonconductor 绝缘体	nonmetal 非金属	nonflammable 不燃的	noncorrodible 防腐的
oct(a)- 八,辛	octane 辛烷	octanol 辛醇	octahedron 八面体	
ortho-(o-) 邻位,正,原	orthoposition 邻位	orthogonal 正交的	orthoacetate 原乙酸酯	
para-(p-) 对位,聚,仲	para-aminophenol 对氨基苯酚	parahydrogen 仲氢	paraldehyde (三)聚乙醛, 仲(乙)醛	
pent(a)- 五	pentahydrate 五水合物	pentagon 五边形,五角形	dinitrogen pentoxide 五氧化二氮	
per- 高,过	perchloric acid 高氯酸	potassium permanganate 高锰酸钾	hydrogen peroxide 过氧化氢	

continued

Prefix (meaning or function)	Examples			
poly- 多,重,聚	polymer	polymorphous	polyethylene	
	聚合物	多形(态)的	聚乙烯	
	polysulfide	polyalcohol		
	多硫化物	多元醇		
pseudo- 伪,假	pseudohalogen	pseudobase	pseudo-first-order	
	假卤素	假碱	准一级	
quadri- 四	quadridentate	quadrielectron	quadrivalent	quadrilateral
	四齿的	四电子	四价的	四边形的
Note: quadribasic acid 四价酸,四元酸				
sub- 在下,次于,低,亚	sub-atomic	subway	subacid	
	亚原子的	地铁	微酸性的	
super-; supra- 上,过,超	supersonic	supraconductivity		
	超声(波)的	超导电性		
sym- 对称的	symcenter	symmetry		
	对称中心	对称现象		
syn-; sys- 共同,与	synchronize	system		
	同步,同时发生	体系		
tetra- 四	tetrad	tetroxide		
	四价	四氧化物		
trans- 反式,转移,越过	trans-2-butene	trans-addition	transferase	transmit
	反-2-丁烯	反式加成	转移酶	传送
tri- 三-	triamine	triaryl methyl	tri-iron tetroxide	
	三胺	三芳基甲基	四氧化三铁	
ultra- 超,过,越	ultraviolet	ultrasonic		
	紫外的	超声速的,超声波的		
under- 次,不足,欠,低,在底下	underground	underproduction		
	地下的	生产不足		
uni- 一,单	unicellular	uniformity		
	单细胞的	均匀,统一,一致		

二、常见的后缀

Suffix (meaning or function)	Examples			
-able 可……的	dependable	changeable		
	可靠的	可变的		
-ase 酶	amylase	esterase	proteinase	
	淀粉酶	酯酶	蛋白酶	
-ate 具有,具有……形状的,(正)盐,酯	nucleate	dentate	chlorate	acetate
	具核的	齿状的	氯酸盐	乙酸盐/酯
	carbonate	nitrate	sulphate	methyl acetate
	碳酸盐	硝酸盐	硫酸盐	乙酸甲酯
-cide 杀	suicide	germicide	pesticide	homicide
	自杀	杀菌剂	杀虫剂	杀人
-ic 表示高价化合物	cupric iodide	ferric bromide	nitric acid	
	碘化铜	溴化铁	硝酸	
-ide ……化物	oxide	hydroxide	sodium hydroxide	
	氧化物	氢氧化物	氢氧化钠	
-ine ……的	marine	porcine	elephantine	
	海的,海产的	猪的	如象的,巨大的	
-ium 元素周期表中金属元素的词尾,有机阳离子的词尾: 鎓	lithium	beryllium	sodium	ammonium
	锂	铍	钠	铵
	oxonium	sulfonium	nitronium	
	氧鎓	锍,硫鎓	硝鎓	
-logy 研究,科学	geology	biology		
	地质学	生物学		
-oid 类似的,相像的	carotenoid	humanoid	bacterioid	
	类胡萝卜素	类人,人形	类菌体	
-ose 糖	glucose	fructose		
	葡萄糖	果糖		
-ous 低价化合物	cuprous iodide	ferrous bromide	nitrous acid	
	碘化亚铜	溴化亚铁	亚硝酸	

continued

Suffix (meaning or function)	Examples			
-philic 亲……的,嗜……的	electrophilic 亲电子的	hydrophilic 亲水的	thermophilic 嗜热的	
-phobic 厌,嫌,疏	hydrophobic 疏水的	acrophobic 恐高的	photophobic 畏光的	
-some 体,易于……的,引起……的	ribosome 核糖体	quarrelsome 喜欢争吵的	troublesome 麻烦的,讨厌的	
-type/-typo/-typ 型,模式	genotype 基因型	phenotype 表型	prototype 原型	atypical 非典型的
-ylene 亚基, —CH_2—词尾	methylene 亚甲基,甲叉	ethylene 乙烯	propyl butylene 丙基丁烯	
-ylidyne 次基, —CH ═词尾	methylidyne 次甲基	ethylidyne 次乙基		

三、常见的词根

Root (meaning or function)	Examples		
aer(i) 空气,空中	aerify 使气体化	aerate 充气于,使透气,通气	aerated water 汽水
Note: -fy, -ate 作为词尾,有时候表示该词类为动词,如 satisfy, emulsify, classify; calculate, graduate			
aqu 水	aqueous 水的	aquagel 水凝胶	aquarium 水族馆
Note: gymnasium 体育馆,健身房(-ium 有时表示金属元素词尾、馆)			
bio 生命,生物	biology 生物学	antibiotic 抗生素	
card(i) 心	cardinal 主要的	cardiac 心脏的	cardiogram 心电图
chrom(o) 颜色、色谱(的词头)	chromatography 色谱法	chromosome 染色体	thin layer chromatography TLC/薄层色谱法
Note: chrome 铬,chrome oxide (三)氧化(二)铬; chromate 铬酸盐			

continued

Root (meaning or function)	Examples					
deca; deka 十,十倍	decameter 十米	dekametre 十米	decade 十年			
gen/gn 出生	genius 天赋	generate 产生	congenate 同源的;同族的			
grad, gress 走,步	gradual 逐步的	degrade 使降级,堕落	digress 离题			
herb 草	herbal 药草的	herbicide 除草剂	herbivore 食草动物			
hydr 水,液,流体;氢,含氢的	hydrous 含水的	hydrolysis 水解	hydrocarbon 烃	carbohydrate 糖类, 碳水化合物		
-ine 人, 女人;状态,药物(素/星)等;表示化学单质	libertine 放荡者	heroine 女主角	Penicillin 青霉素	cephalosporins 头孢菌素		
	tetracyclines 四环素	ofloxacin 氧氟沙星	fluorine 氟	chlorine 氯	bromine 溴	iodine 碘
-ite 亚某酸盐, 亚某酸根	nitrite 亚硝酸根	sodium sulfite 亚硫酸钠				
mar 海,战争	maritime 海上的	martial 军事的	submarine 潜艇			
meter 测量,计,表	thermometer 温度计	photometer 光度计	hydrometer 液体比重计			
not 知道	annotate 注释	notice 告示,通知	notation 记号			
opt 选择,眼睛	adopt 采取,选定	optant 选择者	optical 眼的,视觉的			
photo- 光,感光的	photometer 光度计	photography 图片	photic 感光的	photochemistry 光化学		
phosph- 磷	phosphoric acid 磷酸	sodium phosphate 磷酸钠	dihydrogenphosphate 磷酸二氢根			
pos 放	oppose 反对	dispose 处理	expose 暴露			

continued

Root (meaning or function)	Examples			
-scope 镜，用于观看的器械	microscope 显微镜	telescope 望远镜		
sect 切割	bisect 平分	dissect 解剖	transect 横断	
stereo- 立体	stereochemistry 立体化学	stereoisomer 立体异构体		
sulf/sulph 硫	sulfur trioxide 三氧化硫	sulfite ion 亚硫酸根	sulfate ion 硫酸根	hydrogen sulfide 硫化氢
therm(o) 热	thermometer 温度计	thermodynamics 热力学		

四、派生法在制药及相关专业词汇构词中的应用

1. 表示数目词头的拉丁前缀或希腊前缀

数字	one/1	two/2	three/3	four/4	five/5
天干	甲	乙	丙	丁	戊
前缀	mono-	di-/bi-	tri-	tetra-/quadri-	
前缀	meth-	eth-	prop-	but(a)-	pent(a)-
数字	six/6	seven/7	eight/8	nine/9	ten/10
天干	己	庚	辛	壬	癸
前缀	hex(a)-	hept(a)-	oct(a)-	non(a)-	dec(a)-
数字	eleven/11	twelve/12	thirteen/13	fourteen/14	fifteen/15
前缀	undec-	dodec-	tridec-	tetradec-	pentadec-

2. 表示有机物相应词类的词尾

词尾	有机物类别		构词规律	举例	
-ane	alkane	烷烃	用表示分子中碳原子数的前缀，加上相应有机物类别的词尾，表示不同烃类化合物	methane	甲烷
-ene	alkene	烯烃		ethane	乙烯
-yne	alkyne	炔烃		propyne	丙炔
-enyne	alkenyne	烯炔		1-buten-3-yne	1-丁烯-3-炔
-yl	alkyl	基，烷基	用“yl”取代不同有机物类别词尾的“e”，表示取代基相应类别	pentyl	戊基
-enyl	alkenyl	烯基		propenyl	丙烯基
-ynyl	alkynyl	炔基		ethynyl	乙炔基
-ol	alcohol	醇	将烷烃词尾“ane”中的“e”去掉，分别加上上述词尾，即构成饱和醇、醛、酮	hexanol	己醇
-al	aldehyde	醛		ethanal	乙醛
-one	ketone	酮		2-pentanone	2-戊酮

Note: 也可以用烷基加“aldehyde”表示相应的醛，如 ethyl aldehyde 乙醛；用烃基加烃基再加“ketone”表示相应的酮，如 methyl *n*-propyl ketone 甲基正丙基酮（即 2-戊酮）

ether	ether	醚	两个烃基加“ether”，或用烷氧基加烃名表示醚	*sec*-butyl methyl ether	仲丁基甲醚
				diethyl ether	乙醚
				2-methoxybutane	2-甲氧基丁烷
amine	amines	胺	一个烃基加“amine”是伯氨；两个羟基加“amine”是仲氨；三个羟基加“amine”是叔氨	ethyl amine	乙胺
				phenyl methyl amine	苯甲胺
				N, N-dimethyl propylamine	N, N-二甲基丙胺
-imine	imines	亚胺	含有一个亚氨基时，将烃英文名中“e”换成“imine”；若含有两个以上亚氨基，则在烃名后加数词加“imine”	1-butanimine	1-丁亚胺
				methylenimine	甲亚胺
				polyethylene imine	聚乙烯亚胺
				1, 4-cylcohexanedi-imine	1, 4-环己烷二亚胺
oxide	oxide	氧化物	金属英文名称在“oxide”前面	carbon dioxide	二氧化碳
				zinc oxide	氧化锌

3. 烃类化合物的专业英语词汇常用表达方式

(1) 带有支链的烷烃中常使用的词缀

normal, *n*- 正　　iso- 异　　neo- 新

n-butane 正丁烷　　isobutane 异丁烷　　neopentane 新戊烷

(2) 未取代的烷基常使用的词缀

primary	伯	primary alcohol 伯醇
secondary, *sec-*	仲	*sec*-butyl 仲丁基
tertiary, *tert-*	叔	*tert*-butyl 叔丁基
quaternary	季	quaternary ammonium salt 季铵盐

(3) 结构复杂的烷烃的英文词汇构词规律

支链超过两条时,按基团字母顺序排列支链。例如:

3-ethyl-2-methylhexane	3-乙基-2-甲基己烷
2, 3, 5-trimethyl-4-propylheptane	2,3,5-三甲基-4-丙基庚烷
4-isobutyl-2, 5-dimethylheptane	4-异丁基-2,5-二甲基庚烷

如果是多元炔烃、烯烃等化合物词汇,在其类别词根前加上相应的数词即可。例如:

1, 2, 3-butatriene	1,2,3-丁三烯
1, 3-butadiyne	1,3-丁二炔

环烃、环烃基命名时在相应的开链烃、烃基名前加上词头"cyclo-"。例如:

cyclopropane	环丙烷
3-ethyl-1-methylcyclopentane	3-乙基-1-甲基环戊烷
1-cyclobutylpentane	1-环丁基戊烷
1, 2-dicyclohexylethane	1,2-二环己基乙烷

(4) 芳烃、芳基英文词汇构词规律

aromatic hydrocabons/arenes	芳烃	aryl	芳基
benzene/phene	苯	phenyl	苯基
benzyl	苄基	phenol	苯酚
naphthalene	萘	anthracene	蒽
2-phenyl-2-butene	2-苯基-2-丁烯	phenanthrene	菲
isopropenylbenzene	异丙烯苯	pentylbenzene	戊苯

Note: 两个烷基取代的苯环可用"邻 ortho-,间 meta-,对 para-"表示。如 1,2-二甲苯的英文名称为 1,2-dimethylbenzene 或 *o*-dimethylbenzene.

4. 卤族元素常用的英文词汇构词规律

卤素	卤代基	卤化物	卤酸盐
fluorine 氟	fluoro-氟代,氟基	fluoride 氟化物	
chlorine 氯	chloro-氯代,氯基	chloride 氯化物	chlorate 氯酸盐
bromine 溴	bromo- 溴代,溴基	bromide 溴化物	bromate 溴酸盐
iodine 碘	iodo- 碘代,碘基	Iodide 碘化物	iodate 碘酸盐

o-dibromobenzene	chloroethylene	hexachlorobenzene
邻二溴苯	氯乙烯	六氯代苯

trifluoromethane	1-iodopropane	chloroform
三氟甲烷	1-碘丙烷	氯仿

Note: 卤素取代基命名法(substitutive nomenclature)使用广泛,即在母体化合物前加上相应的卤代基。

benzene hexachloride	六氯化苯
benzyl bromide	溴化苄

Note: 卤化物命名时常用官能团命名法,"化"字表示卤素与某一分子化合。

5. 羧酸及其衍生物英文词汇构词规律

(1) 链状一元羧酸

将英文烃名的最后一个字母"e"去掉,再加上词尾"-oic acid"即可。例如:

methanoic acid	甲酸	formic acid	蚁酸
propanoic acid	丙酸	acetic acid	醋酸
ethanoic acid	乙酸	2-butenoic acid	2-丁烯酸
4-(5-methyl-3-pyridyl) pentanoic acid		4-(5-甲基-3-吡啶基)戊酸	

(2) 羧基连在环烃上的羧酸

一般脂环、桥环、杂环和联环烃的羧酸的构词方式为:可直接在烃英文名后接"carboxylic acid"。例如:

4-methylbenzenecarboxylic acid	4-甲基苯甲酸
cyclohexanecarboxylic acid	环己烷甲酸
2-furancarboxylic acid	2-呋喃甲酸

(3) 羧酸酯、盐

羧酸酯,A 酸 B 酯,构词方式为:醇 B 的烃基英文名在前,A 酸英文名的词尾"-(o)ic acid"改为"-ate"在后即可。例如:

methyl butanoate	丁酸甲酯
benzyl propanoate	丙酸苄酯

羧酸盐,A 酸 B,构词方式是:金属 B 英文名在前,A 酸英文名的词尾"-(o)ic acid"改为"-ate"在后即可。

无机盐,A 化 B 或 A 酸 B,构词方式是:金属 B 英文名在前,后面加上① 卤素词根,再加上某化物后缀"-ide",即为 A 化 B;② 无机酸词根,再加上某酸盐后缀"-ate",即为 A 酸 B。例如:

lithium propanoate	丙酸锂	copper acetate	醋酸铜
sodium chloride	氯化钠	potassium bromide	溴化钾
calcium carbonate	碳酸钙	sodium sulphate	硫酸钠

延伸:醇或酚羟基的氢被金属离子取代所得化合物的构词方式为在醇或酚词尾"-ol"后面,直接加上"-ate"即可。例如:

sodium isopropanolate	异丙醇钠
sodium ethanolate	乙醇钠

(4) 酸酐

在羧酸英文名中,将羧酸词尾“acid”换成“anhydride”即为酸酐英文名称;混合酸酐英文名是将两个不同酸按照首字母顺序写出,省略“acid”,加上“anhydride”即可。例如:

acetic anhydride	醋酸酐	benzoic anhydride	苯甲酸酐
propenoic anhydride	丙烯酸酐	acetic benzoic anhydride	苯乙酸酐

(5) 酰基、酰卤

酰基连在直链烃时,将烃英文名的最后一个字母“e”去掉,加上“-oyl”即为酰基英文名称;当酰基连在环烃上时,直接在烃英文名后加上“carbonyl”。

酰卤构词为:在酰基英文名后直接加上相应的 halide 即可。例如:

ethanoyl	乙酰基	3-cyclohexenecarbonyl	3-环己烯羰基
propanoyl	丙酰基	cyclopropanecarbonyl chloride	环丙基甲酰氯
formyl bromide	溴甲酰	benzoyl iodide	苯甲酰碘

(6) 酰胺

碳链主链包含了酰基碳时,将烃英文名后的“e”去掉,加上“amide”即为酰胺英文名;如果是多酰胺,不需要去掉“e”。当酰基直接连接在环烃上时,在环烃英文名后直接加上“carboxamide”即可。例如:

butanamide	丁酰胺	ethanamide	乙酰胺
hexanediamide	己二酰胺	cyclohexanecarboxamide	环己甲酰胺

6. 药物名称的专业英语词汇构词规律——INN 使用的部分词干

每一种药物都有一个特定名称,而药物名称通常有三种表达方式:国际非专有名(国际非专利药品名称)、化学名和商品名。其中,国际非专有名的英文是 international non-proprietary names for pharmaceutical substance, 缩写为 INN,是由世界卫生组织(World Health Organization, WHO)审定和制定的名称。INN 又称为通用名,一个药物只有一个药品通用名。INN 通常指有活性的药物物质,而不是最终的药品。目前,INN 名称已经被世界各国采用,也是药典中使用的名称。在 INN 中,具有相似药物作用的药物都有相同的词干,表明它们是同类药物。

在 WHO 的国际非专有名的基础上,国家药典委员会出版了《中国药品通用名称(Chinese Approved Drug Names, CADN)》,成为中国药品命名的依据。CADN 中药物的 INN 中文译名是根据 INN 英文名称、药品性质和化学结构以及药理作用等特点,采用音译为主、意译、音意合译或其他译名,尽量与英文名称对应。长音节可简缩,中文名尽可能不多于 5~6 个字,易于发音。

以 procaine 为例,其国际非专有名(INN)是 procaine,通用名(CADN)是普鲁卡因(音译),其词干-caine 音译为卡因,含有此词干的药品为局部麻醉药,如 tetracaine, lidocaine 分别译为丁卡因(音译合译)、利多卡因(音译),均为局部麻醉药。下面列举了部分 INN 常用词干,并进行了举例说明。

词干		药物举例		备注
英文	中文	INN	通用名	
-afenone	-非农(酮)	propafenone etafenone	普罗帕酮 依他苯酮	抗心律失常药
-anserin	-色林	mianserin blonanserin	米安色林 布南色林	5-羟色胺受体拮抗剂
-arabine/-abine	-拉滨 -他滨	fludarabine capecitabine	氟达拉滨 卡培他滨	抗肿瘤药
-arit	-扎利	lobenzarit clobuzarit	氯苯扎利 氯丁扎利	解热镇痛药
-arol	-香豆素	dicoumarol acenocoumarol	双香豆素 醋硝香豆素	抗凝药
-ast	-司特	acitazanolast	阿扎司特	抗过敏药
-azolam	-唑仑	estazolam mexazolam	艾司唑仑 美沙唑仑	镇静催眠药
-azenil	-西尼	flumazenil sarmazenil	氟马西尼 沙马西尼	苯二氮䓬类受体拮抗剂
-azepam	-西泮	diazepam quazepam	地西泮 夸西泮	镇静催眠药
-azocine	-佐辛	pentazocine dezocine	喷他佐辛 地佐辛	镇痛药
-azoline	-唑啉	antazoline tolazoline	安他唑啉 妥拉唑林	抗组胺药或血管收缩药
-azosin	-唑嗪	terazosin prazosin	特拉唑嗪 哌唑嗪	降压药
-bactam	-巴坦	sulbactam tazobactam	舒巴坦 他唑巴坦	β-内酰胺酶抑制剂
-bendazole	-苯达唑	albendazole oxibendazole	阿苯达唑 奥苯达唑	驱肠虫药
-barbital	-巴比妥	phenobarbital hexobarbital	苯巴比妥 海索比妥	镇静催眠药
-bradine	-雷定	cilobradine	西洛雷定	抗心律失常药
-bufen	-布芬	fenbufen butibufen	芬布芬 布替布芬	非甾体抗炎药
-butazone /-buzone	-泰松 /-布宗	phenylbutazone oxyphenbutazone	保泰松 羟布宗	非甾体抗炎药
-caine	-卡因	procaine lidocaine	普鲁卡因 利多卡因	局部麻醉药
-carnil	-卡奈	abecarnil	阿贝卡奈	GABAA 受体激动剂

continued

词干		药物举例		备注
英文	中文	INN	通用名	
cef-	头孢	cefalexin cefixime	头孢氨苄 头孢克肟	头孢菌素类抗生素
-cic	-西克	alonacic	阿洛西克	保肝药
-cillin	-西林	amoxicillin ampicillin	阿莫西林 氨苄西林	青霉素类抗生素
-citabine	-他滨	ancitabine	安西他滨	抗代谢药
-conazole	-康唑	fluconazole miconazole	氟康唑 咪康唑	抗真菌药
-coxib(x)	-昔布	celecoxib imrecoxib	塞来昔布 艾瑞昔布	选择性 Cox-2 抑制剂(非甾体抗炎药)
-cromil	-罗米	probicromil	普克罗米	色氨酸类抗过敏药
-curium	-库铵	atracurium	阿曲库铵	神经肌肉阻断剂
-cycline	-环素	tetracycline minocycline	四环素 米诺环素	抗生素
-dan	-旦/丹	pimobendan	匹莫苯丹	强心药
-dapsone	-(氨)苯砜	acedapsone dapsone	醋氨苯砜 氨苯砜	抗麻风药
-darone	-达隆	amiodarone	胺碘酮	抗心律失常药
-dil	-地尔	minoxidil stevaladil	米诺地尔 甾伐地尔	血管舒张药
-dilol	-地洛	carvedilol	卡维地洛	血管舒张药
-dipine	-地平	nifedipine nitrendipine	硝苯地平 尼群地平	(二氢吡啶类)钙通道阻滞剂
-dopa	-多巴	levodopa carbidopa	左旋多巴 卡比多巴	抗帕金森病药
-ectin	-克丁	abamectin	阿巴克丁	抗寄生虫药
-eridine	-利定	anileridine morpheridine	阿尼利定 吗哌利定	镇痛药
-etanide	-他尼	bumetanide	布美他尼	利尿药
-fenamate	-芬那酯	etofenamate	依托芬那酯	非甾体抗炎药
-fenamic acid	-芬那酸	clofenamic acid	氯芬那酸	非甾体抗炎药
-fentanil	-芬太尼	alfentanil sufentanil	阿芬太尼 舒芬太尼	镇痛药
-fibrate	-贝特	beclobrate fenofibrate	苄氯贝特 非诺贝特	降血脂药

continued

词干		药物举例		备注
英文	中文	INN	通用名	
-flurane	-氟烷	enflurane	恩氟烷	吸入麻醉药
-fungin	-芬净	caspofungin	卡泊芬净	抗真菌类抗生素
-fylline	-茶碱	doxofylline	多索茶碱	平喘药
-giline	-吉兰	mofegiline selegiline	莫非吉兰 司来吉兰	B 型单胺氧化酶抑制剂
-gillin	-洁林	mitogillin	米托洁林	曲霉菌属抗生素
gli-	格列	gliquidone glibenclamide	格列喹酮 格列苯脲	磺酰胺类降血糖药 胰岛素分泌促进剂
guan-	胍-	guanethidini	胍乙啶	抗高血压药
-kacin	-卡星	amikacin butikacin	阿米卡星 布替卡星	氨基糖苷类抗生素
-kalim	-卡林	iptakalim	埃他卡林	钾通道激活药
-mantadine	-曼定/金刚	amantadine	金刚烷胺	抗病毒药
-metacin	-美辛	indomethacin zidometacin	吲哚美辛 齐多美辛	消炎镇痛药
-micin	-米星	etimicin	依替米星	氨基糖苷类抗生素
-monam	-莫南	tigemonam	替吉莫南	内酰胺类抗生素
-motine	-莫汀	famotine memotine	法莫汀 美莫汀	(喹啉衍生物类)抗病毒药
-moxin	-莫辛	octamoxin	奥他莫辛	单胺氧化酶抑制剂
-mycin	-霉素	telithromycin streptomycin	泰利霉素 链霉素	抗生素
nal-	纳	naloxone	纳洛酮	麻醉拮抗剂
-nidazole	-硝唑	metronidazole	甲硝唑	抗真菌药
nifur-	硝呋-	nifuradene	硝呋拉定	抗菌药
-nixin	-尼辛	butanixin	丁尼辛	(苯胺烟酸类)消炎镇痛药
-olol	-洛尔	propranolol bisoprolol	普萘洛尔 比索洛尔	β 受体拮抗剂
-orphan	-啡烷	levallorphan	左洛啡烷	阿片拮抗药
-oxacin	-沙星	norfloxacin ofloxacin	诺氟沙星 氧氟沙星	抗菌药
-oxef	-氧头孢	flomoxef	氟氧头孢	抗菌药
-oxetine	-西汀	fluoxetine paroxetine	氟西汀 帕罗西汀	抗抑郁药

continued

词干		药物举例		备注
英文	中文	INN	通用名	
-oxicam	-昔康	piroxicam meloxicam	吡罗昔康 美洛昔康	非甾体抗炎药
-pamil	-帕米	verapamil gallopamil	维拉帕米 戈洛帕米	钙通道阻滞剂
-parcin	-帕星	avoparcin	阿伏帕星	糖肽类抗生素
-penem	-培南	imipenem meropenem	亚胺培南 美罗培南	非经典的β-内酰胺类抗生素
-pidem	-吡坦	alpidem zolpidem	阿吡坦 唑吡坦	镇静催眠药
-platin	-铂	cisplatin oxaliplatin	顺铂 奥沙利铂	抗肿瘤药
-pramine	-帕明	desipramine	地昔帕明	抗抑郁药
-prazole	-拉唑	omeprazole lansoprazole	奥美拉唑 兰索拉唑	抗溃疡药(质子泵抑制剂)
-pressin	-加压素	desmopressin	去氨加压素	血管收缩剂
-pride	-必利	itopride mosapride	伊托必利 莫沙比利	促胃动力药
-pril	-普利	captopril delapril	卡托普利 地拉普利	血管紧张素转换酶抑制剂
-prim	-普林	trimethoprim brodimoprim	甲氧苄啶 溴莫普林	抗菌药
-profen	-洛芬	ibuprofen suprofen	布洛芬 舒洛芬	消炎镇痛药
-prost	前列素	carboprost latanoprost	卡前列素 拉坦前列素	激素类药物
-relin	瑞林	gonadorelin	戈那瑞林	多肽激素类
-renone	-利酮 /螺酮	drospirenone spirorenone	屈螺酮 螺利酮	甾体避孕药物 (醛固酮拮抗剂)
rifa-	利福	rifamycin	利福霉素	抗生素
-rinone	-力农	amrinone milrinone	氨力农 米力农	强心药
-rubicin	-(矛)比星	doxorubicin zorubicin	多矛比星 左矛比星	抗肿瘤药
-sartan	沙坦	losartan valsartan	氯沙坦 缬沙坦	ACE Ⅱ受体拮抗剂(降压药)

continued

词干		药物举例		备注
英文	中文	INN	通用名	
-semide	-塞米	furosemide torasemide	呋塞米 托拉塞米	利尿药
-serpine	-舍平	reserpine mefeserpine	利血平 美非舍平	抗高血压药
-setron	-司琼	ondansetron alosetron	昂丹司琼 阿洛司琼	5-HT_3受体拮抗剂(镇吐药)
-steine	-司坦	telmesteine	替美司坦	黏液溶解药
sulfa-	磺胺-	sulfadiazine sulfacetamide	磺胺嘧啶 磺胺醋酰	抗感染药
-tadine	-他定	loratadine azatadine	氯雷他定 阿扎他定	组胺-H1受体拮抗剂
-terol	-特罗	salmeterol	沙美特罗	支气管扩张药
-tiazem	-硫䓬	diltiazem	地尔硫䓬	钙通道阻滞药
-tidine	-替丁	cimetidine ranitidine	西咪替丁 雷尼替丁	组胺-H_2受体拮抗剂,抗溃疡药
-tixene	-噻吨	chlorprothixene tiotixene	氯普噻吨 替奥噻吨	抗精神病药
-tizide	-噻嗪	bemetizide polythiazide	泊利噻嗪 贝美噻嗪	利尿药
-toin	-妥英	phenytoin ethotoin	苯妥英 乙苯妥英	抗癫痫药
-trexate	-曲沙	trimetrexate methotrexate	三甲曲沙 氨甲蝶呤	抗代谢药(叶酸类似物)
-tricin	-曲星	metartricin	美帕曲星	抗生素
-triptyline	-替林	nortriptyline amitriptyline	去甲替林 阿米替林	抗抑郁药
-uridine	-尿苷	doxifluridine broxuridine	去氧氟尿苷 溴尿苷	抗肿瘤药
-vastatin /-stat	-伐他汀	lovastatin pitavastatin	洛伐他汀 匹伐他汀	调血脂药
vin-/-vin-	长春	vincristine vindisine	长春新碱 长春地辛	抗肿瘤药
-vir	-韦	pirodavir tecovirimat	吡罗达韦 特考韦瑞	抗病毒药

continued

词干		药物举例		备注
英文	中文	INN	通用名	
-vudine	-夫定	lamivudine zidovudine	拉米夫定 齐多夫定	逆转录酶抑制剂(抗病毒药)
-zepine	-西平	carbamazepine oxcarbazepine	卡马西平 奥卡西平	三环类抗癫痫药

7. 常见药物剂型的专业英语词汇

aerosols	气雾剂	powders	粉剂
capsules	胶囊	solutions	溶液
creams	霜剂	sponge	海绵剂
elixirs	万能药,酏剂	sprays	喷雾剂
emulsion foams	泡沫乳剂	inhalants	吸入剂
gels	凝胶体	suppositories	栓剂
inserts	插入剂	suspensions	混悬剂
lotions	洗剂	syrups	糖浆
magmas	乳浆剂	tablets	片剂
ointments	软膏剂	transdermal patches	透皮药膏
pastes	糊	troches or lozenges	锭剂
plasters	橡皮膏		

Unit 3
Composition

合成法(composition)也叫作复合法,是指由两个或两个以上的独立单词直接合并在一起构成一个新词的方法。该法是英语新词最古老的构词方法,也是常用且简单的一种方法。科技英语词汇中有许多通过合成法构成的新词,即合成词(也叫作复合词)。合成词是一个独立作用的词语,两个或两个以上组成部分之间是不可分离的,中间不能出现其他词语。其书写形式没有太多限制,可以直接连写在一起,也可以采用连字符"-",也可以各自分开。

合成词的组合形式可以是名词+名词、名词+形容词、名词+动词过去分词、形容词+名词、动词+副词等多种形式。合成词的含义在一定程度而言,并不是简单地将两部分含义相加,而是由其基本语义引申出的含义,具有其自身独特的意义和概念。例如:

名词+形容词	doctor positive	高硫的
	fire-resistant	耐火的
名词+名词	carbon steel	碳钢
	diazo methane	重氮甲烷
	rust-resistance	防锈
名词+动词	sleepwalk	梦游
	pain killer	止痛药
名词+动词过去分词	computer-oriented	研制计算机的
	water-cooled	水冷的
名词+动词现在分词	energy-saving	节能的
动词+副词	makeup	化妆品
	check-up	检查
	cutback	稀释
动词+名词	breakwater	防波堤
动词+代词+副词	pick-me-up	兴奋剂
形容词+名词	atomic weight	原子量
	periodic table	周期表
介词+名词	by-product	副产品
副词+动词	upstart	暴发户
副词+过去分词	well-known	著名的
副词+介词+名词	out-of-door	户外
副词+名词	overburden	过重的负担

Unit 4
Shortening

压缩法(shortening)也叫作缩略法,是指将一些词以缩写其部分字母的形式来代替,或把常用的固定词组缩写为一个词的一种构词法。由该法所得词汇为缩略词(abbreviation, abbr.)。缩略词主要是基于现代英语应用范围的拓展和应用频率的提高而产生的,在新词出现之后,尤其是在科技性的文章中,在专业人员达成共识的情况下,其被广泛应用于行业研究和日常生活之中。

缩略词能够使句型复杂的科技性英语文章准确、高效地表达内容,使文章在特定领域的交流中更为方便。缩略词是科学术语终端化的重要体现,也是科技英语词汇的重要构词元素。

英语缩略构词主要有以下几种方式:① 去掉词的后半部,只留前半部,如 Corporation 变为“Corp”;② 去掉词的前半部,只留后半部,如 airplane 变“plane”;③ 去掉词的部分字母,如 China 变成“CN”;④ 保留短语中各主要词的开头字母(通常也叫英语首字母缩略词,是指将一个英语词组中每个单词的首字母进行缩略组合,成为一个新词,词性一般为名词,通常以大写的形式出现),如 over the counter 变成“OTC”。也可以将压缩法简单分为两种方法:一是将单词删去一些字母;二是只取词头字母。

一、将单词删去一些字母

flu	influenza	流行性感冒
h, hr	hour	小时
lab	laboratory	实验室
Corp.	Corporation	股份公司
Dec.	December	十二月
exam	examination	考试
kilo	kilogram	千克、公斤
yr	year	年
Bldg	Building	大楼
Co. Ltd.	Corporation Limited	有限公司
CN	China	中国
maths	mathematics	数学
gly	glycine	甘氨酸

二、只取词头字母

这种方法在科技英语中比较常用。例如：

缩略词(abbr.)	英文(English)	中文(Chinese)
ACS	American Chemical Society	美国化学会
API	Active Pharmaceutical Ingredient	原料药，活性药物成分
CADD	Computer-Aided Drug Design	计算机辅助药物设计
CADN	Chinese Approved Drug Names	中国药品通用名
COVID-19	Coronavirus Disease 2019	2019 冠状病毒病，新型冠状病毒肺炎
CET	College English Test	大学英语水平考试
CIP	clean-in-place	在线清洗，就地清洗
CS	Colorimetric Solution	比色溶液
EP	European Pharmacopoeia	欧洲药典
EPT	English Proficiency Test	英语水平考试(中国)
EST	English for Science and Technology	科技英语
FDA	U. S. Food and Drug Administration	美国食品与药物管理局
GSP	Good Supply Practice	药品销售管理规范
INN	International Non-Proprietary Names for Pharmaceutical Substance	国际非专利药品名称
NCE	New Chemical Entities	新化合物实体
ppm	parts per million	百万分之一
QSAR	Quantitative Structure-Activity Relationships	定量构效关系
SM	starting material	起始原料，起始物料
SOS	Save Our Souls	呼救信号(国际通用)
TOEFL	Test of English as a Foreign Language	非英语国家英语水平考试
WHO	World Health Organization	世界卫生组织

Unit 5
Blending

将两个词的某一部分合并，或者将一个词加上另一个词的一部分，构成一个新词。这样的英语构词法叫作混成法(blending)，也叫作拼缀法。该法是一个具有多产性的方法，尤其是为了适应现代科学技术的发展，利用该法构成了不少用于科技文章、报纸杂志文章的新词。例如：

smog = smoke + fog	烟+雾→烟雾
lunarnaut = lunar + astronaut	月球+宇航员→月球宇航员
biorhythm = biological + rhythm	生物的+节奏→生物节奏
motel = motor + hotel	汽车+旅馆→有汽车停车场的旅馆
positron = positive + electron	正的+电子→正电子
eggwich = egg + sandwich	鸡蛋+三明治→鸡蛋三明治
medicare = medical + care	医学的+照管→医疗保障
sultaine = sulfo + betaine	磺基+甜菜碱→磺基甜菜碱
comsat = communications + satellite	通信+卫星→通信卫星
heliport = helicopter + airport	直升机+飞机场→直升机场
aldehyde = alcohol + dehydrogenation	醇+脱氢→醛(醛是由醇脱氢制得的，并因而得名)

Unit 6
Signs

&	and，和
/	and 或 or，"和"或"或"
M/N	M 和 N，或 M 或 N，两者并取或取一
#	number，号码，#9 = No.9 = number 9
$	dollar，元(美、加等国货币单位)
£	pound，英镑
￥	yuan，元(人民币)
¥	yen，日元
€	euro，欧元

Unit 7

Letters Symbolizing

字母象形构词法(letters symbolizing)的构词模式是“大写字母+连字符+名词”,用以表示事物的外形、产品的型号、牌号等。这些大写字母基本具有摹形状物的作用,让源语读者一看就很容易知道其含义。对译者而言,这些字母为限定词,其含义要与后面的核心词所表达的事物形状相似,翻译时,既要准确表达其科学涵义,又要充分展示其高度显现“象形构词”的特点——“像”。此象形构词法在汉语中屡见不鲜,如“品字形方队、十字街头、八字眉、工字梁”等。

英译汉时,主要采用形译法,亦可根据具体情况具体翻译。例如:

I-bar	工字铁	T-square	丁字尺
U-pipe	U 形管	X-ray	X 射线
V-belt	V 带	U-shaped magnet	马蹄形磁铁

Reading material:

一、科技英语中常见分子式及化学反应式读法

分子式或者化学反应式中的字母读字母音,数字读相应的英文,双键“═”读“double bond”,三键“≡”读“triple bond”,加号“+”读“plus”,箭头“→”读“yields”,正离子“+”读“positive”,负离子“-”读“negative”。例如:

① $C_6H_{12}O_6 \longrightarrow C_2H_5OH + 2CO_2 \uparrow$

读法:C six H twelve O six yields two C two H five OH plus two C O two.

② $NaOH \longrightarrow Na^+ + OH^-$

读法:N sub a O H yields N sub a positive plus O H negative.

二、科技英语中“0”的几种读法

1. 在小数中,如要单个数字读出时,可读成字母音、zero 或 naught。例如:

① 607.08 读“six o(字母音) seven point o(字母音) eight”。

② 0.2 读“zero point two”。

2. 在整数中,一般只读基数词的名称,不读出零。例如:

101 读“one hundred and one”。

3. “0”在表示温度零上零下时，一般读成 zero。例如：

① 10 ℃读“ten degrees above zero centigrade”。

② -3 ℃读“three degrees below zero centigrade”。

4. “0”在号码及年份中，一般读字母音。例如：

① Room 205 读成“room two o(字母音) five”。

② In 1807 读成“in eighteen o(字母音) seven”。

5. “0”在运动会比分中一般读作 zero、nil 或 nothing。例如：

4：0 读成“four to nothing”。

三、GMP 中生产区域的英文名称

生产区域名称		生产区域名称	
英 文	中 文	英 文	中 文
first changing room	一更室	corridor	走廊
hand disinfection room	手消毒室	processing room	加工车间
cleanroom	洁净室	filling room	罐装车间
gowning room	更衣室	immediate package room	内包装车间
airlock room	气闸室	emergency door	安全门

Chapter 7
Writing on the Abstract of a Professional Paper

Unit 1

A Short Guide to Writing Abstracts

What is an abstract

An abstract is a short statement of your paper's contents designed to give its reader a complete, yet concise, idea of what your paper is about. It is a mini-version of your paper.

What types of abstracts are there

There are two types of abstracts that are used in academic writing: informative abstracts and indicative or descriptive abstracts. Each type of abstract serves a purpose. The purpose of the informative abstract is to capsulate a paper. Thus, it is usually this type of abstract that is printed with a paper in professional journals. The indicative or descriptive abstract, on the other hand, is usually only used in conjunction with exceptionally lengthy texts such as entire monographs or broad overviews. This type of abstract does not capsulate the corresponding text (in other words it does not necessarily concern itself with the substance of the paper) and instead only indicates the subject of the text by describing a document's type, the principal subjects covered, and the way the facts are treated (much like a table of contents does for a book). It cannot be considered a substitute for reading a complete paper and is most valuable to librarians. For the purposes of the Massachusetts Public System of Higher Education Conference on Undergraduate Research, Scholarly Creative and Public Service Activities, informative abstracts are needed.

What is the purpose of an abstract

A well-prepared abstract allows a reader to quickly and accurately identify the basic content of your paper, determine its relevance to his or her interests, and finally decide whether or not to read the entire paper. Most readers scan an abstract to see if the related research is of interest to them. The abstract also allows a reader who has little or no interest in the research you present to skip reading the whole document, thus saving time. A well-written abstract is the best means of convincing the right people to read your paper.

What should be in an abstract

A model abstract should contain the following elements: a statement of the purpose of your study, the research methods (or methodology) used to arrive at your results and/or conclusions, the results observed, and the conclusions drawn from your study. These elements will be considered in greater detail in the section below.

Be aware that the elements listed do not necessarily have to be presented in the order shown above. How the elements are sequenced in your abstract depends on the audience for whom the abstract is intended. For instance, if the audience to whom you are writing is exclusively or mainly interested in quickly applying new knowledge, then perhaps you would want to place your most important conclusions and results first, followed by the purpose of the study, methodology, and other-findings and details.

Abstracts in the humanities and social sciences should also contain the above elements. All research, be it in the sciences or the humanities, should have a stated purpose. Research methods in history or sociology may differ substantially from the experimental methods of chemistry, but an abstract, whatever the discipline, must address the methodology of the research. Studies in the humanities and social sciences find results and draw conclusions; these results and conclusions must be included in the corresponding abstract.

How to write an abstract

Many of the following suggestions come from the American National Standard for Writing Abstracts published by the Council of National Library and Information Associations [NOTE: Be aware of the guidelines and conventions of the organization or journal to which you are submitting your abstract and/or research].

The Structure

1. Explain the purpose of your study/paper. This should optimally be only one sentence long. State the primary objectives and scope of the study or the reasons why the document was written (unless these things are already clear from the title of the document or can be derived from the rest of the abstract). Also state the rationale for your research. Why did you do the research? Is the topic you are researching an ignored or newly-discovered one?

2. In terms of methodology (research methods), clearly state the techniques or approaches used in your study. If you want to introduce new methods or approaches in your abstract, keep in mind the need for clarity. For papers concerned with non-experimental work (such as those in the humanities, some social sciences, and the fine arts) describe your sources and your use/interpretation of the sources.

3. Describe your results (the findings of your experimentation), the data collected, and effects-observed as informatively and concisely as possible. These results of course maybe experimental or theoretical, but remember the difference between conjecture and fact and note them in your abstract. Give special priority in your abstract to new and verified events and findings that contradict previous theories. Mention any limits to the accuracy or reliability of

your findings [NOTE: Save the minute details on methods and/or results for the actual paper].

4. By stating your conclusions, you are in essence describing the implications of the results: why are the results of your study important to your field and how do they relate to the purpose of your investigation? Often conclusions are associated with recommendations, suggestions and both rejected and accepted hypotheses.

5. You may wish to include information that is incidental to the main purpose of your paper, but is valuable to those outside your area of study. If you choose to include such information, be careful not to exaggerate its relative importance to the abstracted document. Shown below is an example of how an abstract is to be structured. The components of a well-written abstract are labeled.

Tips for writing abstracts

* Write your paper first, then write the abstract.

* A good abstract should not exceed 250 words. If you can explain your research in 100 words, do not use 200. Remember that each word in print can cost up to $ 0.12.

* Proofread your abstract several times——submit your very best work.

* A good abstract is usually followed by a good paper. The opposite also tends to be true.

* A reader does not want to wade through complicated and unfamiliar terms in the abstract.

* Know your audience and target your abstract accordingly.

* Have a peer read your abstract and then tell you what your research is about. If he or she has difficulty explaining your research, chances are your abstract require revision.

Unit 2

Characteristics and Writing of Abstracts of Scientific and Technical Papers

通常,科技论文都附有中英文摘要。论文摘要的撰写应充分考虑论文的主要内容,应使用简洁的文字将论文的精华高度地概括出来。所以摘要要简明扼要,能独立使用。其中,英文摘要作为科技论文的重要组成部分,有特殊的作用和意义,它是国际上知识传播、学术交流与合作的"桥梁"和媒介,尤其是目前国际上各主要检索机构的数据库对英文摘要的依赖性很强。因此,写好英文摘要对于增加期刊和论文的被检索和引用机会、吸引读者、扩大影响起着不可忽视的作用。

传统的摘要多为一段式 introduction-methods-results and discussion (IMRAD)结构的写作模式,在内容上大致包括引言(introduction)、材料与方法(materials and methods)、结果(results)和讨论(discussion)等主要方面。一般而言,摘要通常与题目、作者及作者所在单位、关键词等一起被收录于 SCI、EI、CA 等索引数据库中,因此,应注意各部分的撰写规则。

一、摘要的目的

有些读者只阅读摘要而不读全文,常根据摘要来判断是否合乎自己需求,有无通读全文的必要。而且大多数检索系统只收录论文的摘要部分或其数据库中只有摘要部分免费提供,因此,摘要应能使读者迅速准确地判明该论文的主要内容。作者应从读者角度出发,在醒目的题目下,精心撰写摘要,让一次文献(原始文献)的读者便于浏览,并且稍经修改,甚至原封不动就能供二次文献(人工检索的文摘、索引、题录、文献卡,以及机读磁带)利用。机检是将摘要直接输入计算机磁带,供全世界读者由国际联机终端,通过卫星通信,向国际上各情报检索系统联机检索,进行文献资料和数据的查询,还可以进一步获得文献全稿。

二、摘要的组成

1. 题目(title)

读者阅读文献时,首先阅读题目,然后考虑是否阅读摘要和全文。索引和摘要服务依赖于准确的题目,便于交叉引用和计算机检索关键词。不恰当的题目很难让相关读者看到,从而使题目完全失去了它应起的作用。因此,题目用词应尽量少,一定要具体而明确,既能概括全篇,又能引人注目。

理想的题目不要过于概括(以致流于空泛、一般化),也不要过于烦琐(无法给予鲜明印象,难于记忆和引证)。写题目时应注意以下几个问题:

(1) 题目是一种"标记",题目用词应尽量少而且要具体而明确。最常采用的形式为名词短语,也可以采用主副标题的形式,很少使用一个完整的句子来描述。句子一般要有主语、谓语、宾语,而题目通常不采用句子形式,要比句子简洁,但词的先后顺序很重要,词序错误是题目写作中常犯的错误。

(2) 下列词语会增加题目的长度,一般在题目中省略不用:"Some Thoughts on","A Few Observations on","Study of","Investigation of","A Final Report on","A Complete Investigation of"等。

(3) 题目中不要使用缩略语、化学分子式、专利商标名称、行话、罕见的或过时的术语。

(4) 按惯例,英语题目第一个词和每个实词的第一个字母要大写。

2. 作者(author)

论文的署名作者应满足下列条件:(1) 必须参与过本项研究的设计和开创工作,如在后期参加工作,必须赞同研究的设计。(2) 必须参加过论文中的某项观察和获取数据工作。(3) 必须参与过对所观察的现象和所取得的数据的解释,并从中得出论文的结论。(4) 必须参加过论文的撰写。仅对这项科研工作给予过支持和帮助,但不符合上述条件者,不能列为作者。

科技文章可以个人署名,也可以用集体名义署名,并写明工作单位和地址,便于联系。国外有些期刊还要求作者写出其最高学位,以示郑重负责。如果作者不止一位,作者名单排列的顺序目前尚没有统一的规则。一般按对论文的贡献大小从前往后排列。

3. 摘要的内容(content)

摘要应反映论文的主要信息,简明而不晦涩难懂,摘要包含如下几个方面,以成果、结论为重点。

(1) 目的——研究的目的及重要性。

(2) 方法——研究的内容、方法和取得的进展,应该详而不繁。

(3) 成果——主要成果及贡献。

(4) 结论——评议论文的价值。

各国最有影响的化学杂志对文摘写作有些共同要求。归纳起来有三条:

(1) 要说明实验或论证中新观察的事实、结论。如有可能,还要说明新的理论、处理、仪器、技术等的要点。

(2) 要说明新化合物的名称,新数据,包括物理常数等。如不能说明,也要提到这些。提到新的项目和新的观察是很重要的,哪怕它们对本论文的主旨来说仅仅是附带而已。

(3) 在说明实验结果时,要说明采用的方法。如是新方法,还应说明其基本原理、操作范围和准确度。

4. 关键词(key words)

按照惯例,摘要的末尾要加关键词,关键词可以是单词,也可以是由专业词汇构成的词组。关键词的主要作用是作为索引和检索系统的补充,应能鲜明而直观地表达该论文的主题内容。每一个关键词都要表征某个或某一方面的确切含义,通过数个关键词的逻辑组合可以准确提示一篇论文的主题。显然,以一个包含较多内容的词组或短语作为关键词是不符合关键词原意与使用要求的。关键词的选择除了要清晰地表征、提示论文主题内容,还要考虑这些关键词是否有助于论文的检索。一篇学术论文关键词选择是否恰当,关系到该论文被检索的概率及该成果的利用率。因此,正确理解关键词的写作与使用,将直接影响论文自身的有效价值。

关键词一般为3~8个,根据"自由词汇表"或根据最新的权威性专业词汇从论文中选用。比如,杂志*European Journal of Medicinal Chemistry*(《欧洲药物化学》)要求投稿时,关键词不能超过6个。

三、摘要的长度

各种不同的刊物对摘要长度的要求不同,限制摘要的具体长度是不可能的。对简短扼要的论文来说,摘要一般要求短些,不超过全文3%比较恰当。但相对来说,短篇论文的摘要与正文的比例,比长篇论文的摘要与正文的比例大些。长的摘要如有分段必要,每段可有100~150个单词,也可短到十几个单词。

博士、硕士论文的摘要有两种:一种放在论文前,与上述学术论文的相同;另一种是文摘,实质上是论文的缩写,供评议、审查之用,可长达3 000个单词。总的来说,摘要必须简明扼要,一般化学、药学等专业论文摘要不超过250字,且应尽量简短精练。比如,药物化学领域著名期刊*Journal of Medicinal Chemistry*(《药物化学》)对摘要做出如下要求:

Articles, Drug Annotations, Perspectives, and Viewpoints must have an abstract following the title page. For Articles, Drug Annotations, and Perspectives, 150 words are usually adequate; for Viewpoints, 50 words are adequate. Abstracts should be presented in a findings-oriented format in which the most important results and conclusions are summarized. Descriptive names or code names may be used in the abstract.

国际顶级刊物*Science*(《科学》)对研究论文摘要的具体要求为:Abstract of Research Articles should explain to the general reader why the research was done, what was found and why the results are important. They should start with some brief BACKGROUND information: a sentence giving a broad introduction to the field comprehensible to the general reader, and then a sentence of more detailed background specific to your study. This should be followed by an explanation of the OBJECTIVES/METHODS and then the RESULTS. The final sentence should outline the main CONCLUSIONS of the study, in terms that will be comprehensible to all our readers. The Abstract is distinct from the main body of the text, and thus should not be the only source of background information critical to understanding the manuscript. Please do not include citations or abbreviations in the Abstract. The abstract should be 125 words or less. For Perspectives and Policy Forums please include a one-sentence abstract.

RSC Medicinal Chemistry（原名 *MedChemComm.*）要求摘要符合下述要求：The abstract is a single paragraph which summarises the findings of your research. It will help readers to decide whether your article is of interest to them.The length can vary from 40 to 150 words, but it should always be concise and easy to read, with recognizable words and phrases.It should set out the objectives of the work, the key findings and why this research is important (compared to other research in its field). It should emphasize (but not overstate) the significance and potential impact of the research in your article.Avoid including detailed information on how the research was carried out. This should be described in the main part of the manuscript. Like your title, make sure you use familiar, searchable terms and keywords.

有的期刊要求提供图文摘要(graphical abstract)，如 *European Journal of Medicinal Chemistry*（《欧洲药物化学》）对图文摘要做出如下要求：

A graphical abstract is mandatory for this journal. It should summarize the contents of the article in a concise, pictorial form designed to capture the attention of a wide readership online. Authors must provide images that clearly represent the work described in the article. Graphical abstracts should be submitted as a separate file in the online submission system. Image size: please provide an image with a minimum of 531×1328 pixels (h×w) or proportionally more. The image should be readable at a size of 5×13 cm using a regular screen resolution of 96 dpi. Preferred file types: TIFF, EPS, PDF or MS Office files.

四、撰写摘要的注意事项

1. 规范

摘要虽然通常在正文前面，但往往是在最后写作的。写之前，应查询拟投稿期刊的读者须知，以了解其对摘要的字数和形式的要求。写摘要要使用正规英语、标准术语，避免使用缩略词。如确有需要(如避免多次重复较长的术语)使用非同行熟知的缩写，应在缩写符号第一次出现时给出其全称。最好用第三人称写作，避免各种时态混用，也避免把陈述式和命令式掺杂使用。

2. 精练

摘要要精练，使用简短的句子。为确保摘要简洁而充分地表述论文的 IMRAD，可适当强调研究中的创新、重要之处(但不要使用评价性语言)，尽量包括论文中的主要论点和重要细节(重要的论证或数据)。不宜列举例证，不宜与其他研究工作对比。用词必须确切，切忌言之无物，衍文赘语连篇。表达要准确、简洁、清楚；注意表述的逻辑性，尽量使用指示性词语表达论文的不同部分(层次)。如使用“We found that...”表示结果，使用“We suggest that...”表示讨论结果的含义等。要力求字斟句酌，言简意赅。

精简英语摘要，要注意如下几方面：

(1) 不用开场白

要避免使用诸如下列开场白：

This is a report on...	这是关于……的一篇报道
In this paper, the author presents...	在这篇论文中，作者介绍……
The purpose of this article is to...	本文目的在于……

应把这类套话简写成：

The paper/author suggests/presents...	本文(作者)提出(介绍)……

(2) 不用标题语

摘要开头处，不可将标题中用过的词语再重复一遍。摘要中用词应为潜在的读者所熟悉，尽量避免使用化学结构式、数学表达式、角标和希腊文等特殊符号以及引用图等非文字性内容，以节省磁带存储空间，便于计算机的输入和输出。若无法回避引用文献，应在引文出现的位置将引文的书目信息标注在方括号内。

(3) 少用形式主语

使用形式主语易引起下列问题：占去句首醒目的强调位置，平添一个 that 主语从句，使重要概念和事实不能及早入目，先睹为快；主句用平淡无力的存在动词 be，而将有表现力的行为动词隐藏在后面的从句中；提出的观点似乎不是出自作者本意，使人怀疑作者论据不足。

例如：

a. It is concluded that a new method has to be devised.

改为：(Conclusion:) A new method has to be devised.

(结论：)不得不想出一个新方法。

b. It will be seen that further research is needed.

改为：Further research is needed.

需要做进一步的研究。

3. 具体

摘要的每个论点都要具体鲜明。一般不笼统地讲论文“与什么有关”，而直接讲论文“说明什么”。例如，不讲“Azo dyes are prepared by the coupling of diazonium salts”(偶氮染料是用重氮盐的偶联方法制备的)，因为读者不能重复你的试验；而讲“Azo dyes are prepared when a phenol or aryl amine is treated with a diazonium salt solution”(偶氮染料是苯酚和芳基胺用一种重氮盐溶液处理而制成的)。

4. 完整

摘要本身要完整。有些读者是利用摘要杂志或索引卡片进行研究工作的，很可能得不到全篇论文，因此注意不要引用论文某节或某张插图来代替说明。

5. 全面

不要过于简单地只把论文标题加以扩展,使读者无法得到全文梗概。这种写法在文献中是有的,要注意避免。

Unit 3
Examples of English Abstracts of Scientific and Technical Papers

示例一

{Abstract}: Evodiamine has many biological activities. Herein, we synthesize 23 disubstituted derivatives of N14-phenyl or the E-ring of evodiamine and conduct systematic structure-activity relationship studies. *In vitro* antiproliferative activity indicated that compounds **F-3** and **F-4** dramatically inhibited the proliferation of Huh7 (IC_{50} = 0.05 or 0.04 μM, respectively) and SK-Hep-1 (IC_{50} = 0.07 or 0.06 μM, respectively) cells. Furthermore, compounds **F-3** and **F-4** could double inhibit topoisomerases I and II, inhibit invasion and migration, block the cell cycle to the G2/M stage, and induce apoptosis as well. Additionally, compounds F-3 and **F-4**could also inhibit the activation of HSC-T6 and reduce the secretion of collagen type I to slow down the progression of liver fibrosis. Most importantly, compound **F-4** (TGI=60.36%) inhibited tumor growth more significantly than the positive drug sorafenib. To sum up, compound **F-4** has excellent potential as a strong candidate for the therapy of hepatocellular carcinoma.

参考文献: LEI F, XIONG Y X, WANG Y Q, et al. Design, synthesis, and biological evaluation of novel evodiamine derivatives as potential antihepatocellular carcinoma agents[J]. Journal of Medicinal Chemistry, 2022, 65 (11): 7975-7992.

示例二

{Abstract}: The natural product oridonin has the potential to be a broad-spectrum antineoplastic agent. To develop oridonin analogues with high potency, a series of novel oridonin analogues were designed and synthesized by removing the multiple hydroxyl groups of parent compound. The representative analogues **14**, **19**, and **26** exhibited potent anticancer effects against K562, MDA-MB-231, SMMC-7721, and MCF-7 cells. Further structural modification on their 14—OH generated more potent derivatives **16n**, **21d**, and **28d** respectively, in which the IC50 value of compound **16n** was 50-fold more potent than parent oridonin in K562 cells. Furthermore, compound **16n** significantly induced the cell cycle arrest of K562 cells at the G2 phase and increased the fraction of apoptotic cells. Importantly, compounds **16n**, **21d**, and **28d** exhibited good antitumor activities in H22 allograft mice *in vivo*. These results suggest that compounds **16n**, **21d**, and **28d** deserve further development as promising candidates for the treatment of cancers.

参考文献：LIU J K, XIE S W, SHAO X, et al. Identification of new potent anticancer derivatives through simplifying the core structure and modification on their 14-hydroxyl group from oridonin[J]. European Journal of Medicinal Chemistry, 2022, 231: 114155.

示例三

{摘要}：本研究设计合成了 28 个全新 12*N* 取代苦豆碱衍生物并测定其在乳腺癌细胞 MDA-MB-231 中下调 PD-L1 水平的活性。其中，化合物 **7f** 具有较高的下调 PD-L1 活性，呈时间和剂量依赖性，且显示出较低的细胞毒性。**7f** 可浓度依赖性地激活共培养 T 细胞对肿瘤细胞的杀伤活性，显示出肿瘤免疫治疗的潜力。进一步研究显示，7f 可能通过溶酶体途径介导 PD-L1 的降解。该研究为苦豆碱类化合物发展为一类全新小分子肿瘤免疫抑制剂提供了有益的指导。

{Abstract}： Totally 28 new 12*N*-substituted aloperine derivatives were designed, synthesized and evaluated for their down-regulating PD-L1 activities in breast cancer MDA-MB -231 cells. Among them, compound **7f** could significantly down-regulate PD - L1 level in concentration-and time-dependent manners, and exhibit a low cyto-toxicity. It activated the killing activity of co-cultured T cells against tumor cells in a concentration-dependent manner, showing the potential of tumor immunotherapy. Further study indicated that **7f** mediated the degradation of PD-L1 through a lysosomal pathway. This study provides useful guidance for the development of aloperine compounds into new small molecule tumor immune suppressants.

参考文献：张昕彤，王坤，曾庆轩，等. 基于 PD-L1 肿瘤免疫苦豆碱类衍生物的合成与活性研究[J]. 药学学报，2022，57(4)：1085-1094.

示例四

{摘要}：以热爆预处理制备的黑木耳发酵原浆为原料，采用热水浸提法提取黑木耳粗多糖(Crude Auricularia heimuer polysaccharide，CAHP)。通过 H_2O_2 氧化水解和超声物理水解联用的方法降解 CAHP，经阴离子交换柱层析和凝胶过滤柱层析分离，获得黑木耳寡糖 1(Auricularia heimuer oligosaccharide 1，AHO1)。单糖组成分析、高 分辨质谱、红外光谱和核磁共振波谱等表征结果表明，AHO1 是以甘露糖为主，由 6 种单糖组成，各糖残基通过 α-糖苷键连接，聚合度为 2~10 的低聚糖功能分析结果表明，AHO1 具有一定的自由基清除能力，且能够有效减弱 H_2O_2 诱导的 HepG2 细胞氧化应激反应。抑菌实验结果显示，AHO1 不仅对大肠杆菌 DH5α 和金黄色葡萄球菌 ATCC6538 具有较好的抑制作用，还能有效抑制带氨苄抗性的基因工程菌活性。

{Abstract}： In this study, Auricularia heimuer fermentation puree prepared by thermal explosion pretreatment was used as raw material. Crude Auricularia heimuer polysaccharide (CAHP) was extracted by hot water extraction. CAHP was degraded by a combination of H_2O_2 oxidative hydrolysis and ultrasonic physical hydrolysis. Then, CAHP was separated by anion exchange column chromatography and gel filtration column chromatography to obtain Auricularia heimuer oligosaccharide 1 (AHO1). Analyzed by monosaccharide composition analysis, high-resolution mass spectrometry, infrared spectroscopy, and nuclear magnetic resonance, AHO1 is an oligosaccharide with a degree of polymerization of 2−10 connected by α-glycosidic bonds, which is composed of six kinds of monosaccharides, maliy mannose. The results of functional analysis showed that AHO1 had certain free radical scavenging ability and could effectively attenuate the oxidative stress response of HepG2 cells induced by H_2O_2. The results of antibacterial experiments showed that AHO1 not only has a good inhibitory effect on *Escherichia coli* DH5α and *Staphylococcus aureus* ATCC6538, but also can effectively inhibit the activity of genetically engineered bacteria with ampicillin resistance.

参考文献:胡皓程,李文利,张嘉宁. 黑木耳寡糖的提取、结构表征及生物活性[J]. 高等学校化学学报, 2021, 42(8): 2465−2473.

Reading Material:

The Structure of a Scientific Research Paper

科学研究论文结构

英文科技论文是国际学术交流必需的载体。为便于快速阅读科技论文的重点内容,首先要掌握英文科技论文的一般结构模式,然后根据读者阅读论文的不同目的有针对性地寻求答案。

英文科技论文与中文科技论文的结构模式基本相近,对于研究报告性的科技论文(scientific research article),其结构包括:题目(title)、作者(authors)、作者所在单位(affiliation)、摘要(abstract)、关键词(key words)、引言(introduction)、材料与方法(materials and methods)、结果与讨论(results and discussion)、结论(conclusion)、致谢(acknowledgements)、引用文献(references)。

摘要部分通常包括研究动机、研究方法、主要结果和简要结论。它起着缩微全文、吸引读者和扩大发行的作用。

引言部分主要是介绍论文中要解决的实际问题是什么,研究该问题的科研背景以及国内外相关研究进展如何等。通过总结、简要概括相关参考文献的主要研究进展来引发读者对该问题的兴趣。一般段落不长,常见一至四个段落。

材料与方法部分主要是介绍实验所使用的试剂、原料和仪器的来源,即试剂的级别、型号、厂家或公司;另外详细介绍实验的具体方法和具体步骤。其目的是便于其他科研人员能够如法重复实验并得到相同的结果。

结果部分主要是表述作者所得到的实验结果,常以表或图的形式加以描述。讨论部分往往紧接在结果部分合二为一,用于解释实验中所得到的重要结果。

结论部分为总结所得结果和假设并就后续研究提出建议。

Vocabulary

词汇表

A

abiraterone	*n.*阿比特龙
abrade	*vt.* 磨,擦,擦伤
absorption	*n* 吸收
accelerated phase	加速期
accumulation	*n.* 积累,积聚物
acetic acid	乙酸
acetone	*n.* 丙酮
acetonitrile	*n.* 乙腈
acetoxy group	乙酰氧基
acidic	*adj.* 酸的,酸性的
acoustophoresis	*n.* 声泳
acumen	*n.* 敏锐,聪明
acyl	*n.* 酰基
acylamino	*n.* 乙酰氨基
adaptive resistance	适应性耐药
adiabatic	*adj.* 绝热的,不传热的
adjuvant therapy	辅助疗法
administration	*n.* 给药
adolescent	*n.* 青少年; *adj.* 青春期的,青少年的
adulterate	*v.* 掺杂,掺假
adverse event	不良反应
aerobic	*adj.* 需氧的,有氧的
affinity	*n.* 亲和力
aggressive systemic mastocytosis	系统性肥大细胞 *n.*增多症
air handling	空气处理

algorithms	*n.* 算法
alkaloid	*n.* 生物碱
alkyl	*n.* 烷基
allele deletion	等位基因缺失
allergic rhinitis	变应(过敏)性鼻炎
alternative therapy	替代疗法
7-aminocephalo-sporanic acid	7-氨基头孢烷酸
6-aminopenicillanic acid	6-氨基青霉烷酸
Ammonium formate	甲酸铵
amoxycillin	*n.* 阿莫西林
amphiphilicity	*n.* 双亲性
Ampicillin	*n.* 氨苄西林
anaesthetics	*n.* 麻醉剂
analogues	*n.* 类似物
analytical	*adj.* 分析的
anhydrous	*adj.* 无水
annulus	*n.* 环(带,形,状)物,环形套筒
antacids	*n.* 抗酸药
antagonists	*n.* 拮抗剂
antiandrogen	*n.* 抗雄激素物质
antibacterial	*adj.* 抗菌的
antibiotic	*adj.* 抗生的, 抗菌的; *n.* 抗生素
antibody	*n.* 抗体
antifoaming	*adj.* 防沫的,消沫的
antifungal	*adj.* 抗真菌的

antimicrobial	*n.* 抗菌剂
antioxidants	*n.* 抗氧剂
antiseptic	*adj.*; *n.* 防腐(消毒,杀菌)的;防腐(抗菌,消毒)剂
anxiety	*n.* 焦虑
apoptosis	*n.* 细胞凋亡
appraisal	*n.* 评价,评估
appropriateness	*n.* 适宜性
arrhythmic	*adj.* 心律不齐的
artificial intelligence	人工智能
aseptic	*n.* 防腐剂; *adj.* 无菌的
asparagine	*n.* 天冬酰胺
augment	*v.* 增大(加,长),扩大(张)
autopsies	*n.*[复]尸体解剖(检验)
Aztreonam	*n.* 氨曲南

B

bacterial	*adj.* 细菌的
base	*n.* 碱
batch	*adj.* 间歇的,分批的; *n.* 一批,批料,批次,批量
beaker	*n.* 烧杯
benzene	*n.* 苯
bicyclic	*adj.* 双环的
bioavailability	*n.* 生物利用度
bioequivalent	*adj.* 生物等效的
blast crisis	*n.* 急变期
blotting	*n.* 印迹
botanical	*n.* 植物性药材; *adj.* 植物学的,来自植物的
breakdown	*n.* 数字明细
breast cancer	乳腺癌
buffer	*vt.* 缓冲; *n.* 缓冲剂
bunsen burner	本生灯
by-product	*n.* 副产物

C

caffeine	*n.* 咖啡因,咖啡碱
calcitonin	*n.* 降钙素
calcium carbonate	碳酸钙
calibrate	*v.* 校准
capacity	*n.* 产能
carboxyl group	羧基
carboxylic acid	羧酸
carcinogenicity	*n.* 致癌性,致癌作用
cardiologists	*n.* 心内科医生
cardiovascular	*adj.* 心血管的
carry-over	(样品)遗留[如从上次实验中残留至下次实验]
catechin	*n.* 儿茶酸
cavitation	*n.* 空泡,成腔,空化
cavity	*n.* 洞,腔
Ceftazidime	*n.* 头孢他啶
cephalomannine	*n.* 三尖杉宁碱
cephalosporium	*n.* 假头状孢子头
characterization	*n.* 表征
charcoal	*n.* 木炭
charge	*n.* 电荷
chemotherapy	*n.* 化疗
chloride	*n.* 氯化物
chloroform	*n.* 氯仿
chlorophyll	*n.* 叶绿素
chloroquine	*n.* 氯喹
cholesterol	*n.* 胆固醇
chromatogram	*n.* 色谱,色谱图
chromosome	*n.* 染色体
chronic	*adj.* 慢性的, 习惯性的
chronically	*adv.* 慢性地
Ciprofloxacin	*n.* 环丙沙星
cisplatin	*n.* 顺铂
clavulanic acid	克拉维酸
clinical	*adj.* 临床的,病房用的
clinical disciplines	临床学科
clinical pharmacy	临床药学
clinical symptom	临床症状
clinical trial	临床试验
clinically	*adv.* 临床上地
clinician	*n.* 临诊医师,门诊医师
clog	*n.* 堵塞,阻塞

co-administrated drugs	联合用药
coating	n. 包衣
cohesion	n. 内聚力
colligative	adj. 综合的,概括的
colony	n.(生物)群体
column resins	树脂柱
combinatorial chemistry	组合化学
compendial	n. 药典
condenser	n. 冷凝器
confidence	n. 信任,自信心
conformational	adj. 构象的
conformation	n. 构象
conjunctivitis	n. 结膜炎
consistency	n. 一致性
contaminant	n. 污染物
contamination	n. 玷污,污染,污染物
contraindication	n. 禁忌证
corn steep liquor	玉米浸液
criteria	n. 标准,规范,尺度
crospovidone	n. 交聚维酮
crystallization	n. 结晶
cycle	n. 循环
cytopenia	n. 血细胞减少症
cytoplasm	n. 细胞质
cytotoxic	n. 细胞毒性

D

data deluge	数据洪流
decimal	n. 小数; adj. 十进位的, 小数的
dedication	n. 贡献,奉献
degradability	n. 降解性
degradation	n. 降解
delineate	vt. 描外形,刻画,描绘
delivery systems	给(递)药系统
demographic	adj. 人口统计学的
dendrimers	n. 树枝状聚合物
depict	vt. 描述,描写
depression	n. 忧郁
derivative	n. 衍生物;adj. 衍生的
dermal	adj. 皮肤的
dermatologists	n. 皮肤科医师
design of the facility	(厂房)设施设计
design specification	设计规范
designation	n. 认定
detector	n. 检测器
detergent	n. 洗涤剂
deviation	n. 偏差
diagnose	v. 诊断
diarrhea	n. 痢疾,腹泻
dietary supplement	膳食补充剂
digalloyl group	鞣酰基
diluent	n. 稀释剂
dilution	n. 稀释,稀释法
dissociate	vt. 使分离,使离解
distillation	n. 蒸馏
distillation flask	蒸馏瓶
diuretic	n. 利尿剂; adj. 利尿的
dopamine	n. 多巴胺(一种神经递质)
dose modification	剂量调整
drift	n. 漂移
drug resistance	耐药性
drug uptake	药物吸收
druggability	n. 成药性
duration	n. 持续时间,用药期限
dynamics	n. 动力学
dysfunctions	n. 功能失调
dysregulation	n. 异常调控

E

ebullating-bed	沸腾床
edema	n. 浮肿,水肿
electrocardiographic	adj. 心电图的
eluent	n. 洗脱液
emulsion	n. 乳剂
enantiomer	n. 对映异构体
encapsulation	n. 封装
encrustation	n. 垢体

endocrine	*n.* 内分泌
endocyclic	*adj.* 桥环的
endogenous	*adj.* 内源的，内源性的
endothermic	*adj.* 吸热的，内热的
endotoxin	*n.* 内毒素
Enoxacin	*n.* 依诺沙星
entity tumors	实体瘤
entrainment	*n.* 挟带，夹带，带去，传输
entropic	*n.* 熵
enzyme	*n.* 酶
Escherichia coli	大肠杆菌
esterify	*v.*（使）酯化
esters	*n.* 酯
ether	*n.* 醚
excretion	*n.* 排泄
exosome	*n.* 外泌体
exothermic	*adj.* 放热的，发热的
exposure	*n.* 暴露量
extemporaneous	*adj.* 即席的，临时的
extract	*vt.* 榨出，萃取，提取，蒸馏（出）
extraction	*n.* 萃取，提取
extraneous	*adj.* 外源性，外来的

F

facility	*n.* 设备
feedstuff	*n.* 饲料
fermentation	*n.* 发酵
fibrosis	*n.* 纤维化
filing	*n.* 备案
filter paper	滤纸
filtrate	*n.* 滤（出）液 *v.* 过滤
finance	*n.* 金融
finished product	制剂产品
flavonoid pigment	黄酮类颜料
flexible	*adj.* 灵活的
fluid	*n.* 流体
fluorine	*n.* 氟
formic acid	甲酸
formulation	*n.* 制剂
fragment	*n.* 片段
frequency	*n.* 频率
fungal = fungous	*adj.* 真菌的
fungus	*n.* 真菌（包括霉菌、酵母菌和伞菌等）
fused ring	稠环
fusion	*n.* 融合

G

gall	*n.* 胆汁
gallic acid	五倍子酸
gastrointestinal perforation	胃肠道穿孔
gastrointestinal stromal tumors	胃肠道间质瘤
gelatin/gelatin	*n.* 骨胶，明胶
gelatine	*n.* 明胶
Gemcitabine	*n.* 吉西他滨
genomic	*adj.* 基因组的
genotoxic	*n.* 基因毒性，遗传毒性
geriatric population	老年患者
germicidal	*adj.* 杀菌的
glucose	*n.* 葡萄糖
granulation	*n.* 造粒
gravimetric	*adj.* 重量分析的，重量的
guide	*n.* 指南

H

hallmark	*n.* 检验印迹
hazard	*n.* 危害物
healthcare	*n.* 医疗卫生服务
helicase	*n.* 解螺旋酶
heparin	*n.* 肝素
herbal	*adj.* 草药的
hereditary	*adj.* 遗传的
heterocyclic	*adj.* 杂环的
heterogeneous	*adj.* 不均匀的，多相的
hiatus	*n.* 脱落，裂缝
high-throughput screening	高通量筛选

Hirsch funnel	赫尔什漏斗
holding	*n.* 贮藏
homogeneous	*adj.* 均一的,均相的,均匀的,同质的
homology	同源, 同源性
hormone	*n.* 激素,荷尔蒙
hydrogen	*n.* 氢
hydrolyzable	*adj.* 可水解的
hydrolyze	*vi.* 水解
hydrophobic	*adj.* 疏水的
hydroxyl group	羟基
hygienic	*adj.* 卫生学的,卫生的
hypothyroidism	甲状腺功能减退

I

ice water bath	冰水浴
identification	*n.* 鉴定,确认
Imatinib	*n.* 伊马替尼
immobilize	*vt.* 固定化
immunomodulatory	*n.* 免疫调节
implantation	*n.* 对端植入法
impurity	*n.* 杂质
in vitro	体外,在试管内
in vivo	体内,在活的有机体内
incorporate	*v.* (使)合并(并加)
indication	*n.* 适应证
individual	*n.* 个体
infectious	*adj.* 有传染性的,易传染的
infective	*n.* 感染
inferential	*adj.* 推论的,推理的
infrared	*n.* 红外
ingredient	*n.* 成分,因素
inhalation	*n.* 吸入
inhalational	*adj.* 吸入的
inhale	*v.* 吸入
inhibitor	*n.* 抑制剂
in-house	*n.* 内部
initial	*adj.* 最初的
injection	*n.* 注射
innate resistance	天然耐药
insect	*n.* 昆虫
instantaneous	*adj.* 瞬间的,即刻的
insulin	*n.* 胰岛素
integrity	*n.* 完整性
interchangeable	*adj.* 可交换的
interferon	*n.* 干扰素
interferon-alpha	*n.* α-干扰素
intermediate	*n.* 中间体
intermittent	*adj.* 间歇性的
intramuscularly	*adj.* 肌内(地),肌内(注射)
intravenous	*adj.* 静脉内的
intriguing	*adj.* 引起兴趣(或好奇心)的,有迷惑力的
invasive	*adj.* 侵略的,侵害的,侵袭的
ionisation	*n.* 电离
ionizing radiation	电离辐射
Irinotecan	*n.* 伊立替康
irrigate	*vt.* 冲洗
isolation	*n.* 分离
isopenicillin	*n.* 异青霉素
isopropyl	*n.* 异丙基

J

jurisdiction	*n.* 司法权, 管辖权

K

kiln	*n.*(砖, 瓦)窑, (火)炉, 干燥器; *vt.* (窑内)烘干,窑烧

L

laminar	*adj.* 层(式, 状, 流)的,分层的,片(状)的
lethal	*adj.* 致死的
leukemia	*n.* 白血病
levofloxacin	*n.* 左氧氟沙星
life-expectancy	预期寿命
ligand	*n.* 配体
ligroin	*n.* 轻石油,石油英,

	粗汽油
lipid	*n.* 脂质
lipophilicity	*n.* 亲脂性
liposomes	*n.* 脂质体
local tissue	局部组织
logic	*n.* 逻辑,思维方式
lot	*n.* 批次
lymph	*n.* 淋巴,淋巴液
lymphoblastic	*adj.* 成淋巴细胞的
Lymphoma	*n.* 淋巴瘤
lyophilize	*n.* 低压冻干法,升华干燥
lysis/[复] lyses	*n.*(病的)渐退,消散,细胞溶解

M

macrocyclic	*adj.* 大环的
magnesium stearate	硬脂酸镁
malignant	*adj.* 有恶意的,恶毒的恶性的
malignant tumors	恶性肿瘤
mammalian	*n.* 哺乳动物
manipulation	*n.* 操作
manufacture	*n.* 制造
manufacturer	*n.* 制造商
margin	*n.* 界值
marketing authorisation	上市许可
mass balance	质量守恒
material	*n.* 材料
matrix	*n.* 基体; 基质
mechanism	*n.* 机制
medical device	医疗器械(设备)
medical records	病例档案
medicinal chemistry	药物化学
melon	*n.*(各种的)瓜
melting points	熔点
memorabilia	[复] *n.* 大事记,值得纪念的事
mercaptopurine	*n.* 巯基嘌呤
mesilate	*n.* 甲磺酸
mesylate	*n.* 甲磺酸盐
metabolism	*n.* 新陈代谢
metabolite	*n.* 代谢产物
metastatic	*adj.* 转移性的
dermatofibrosarcoma protuberans	隆突性皮肤纤维肉瘤
methanol	*n.* 甲醇
metrology	*n.* 计量学
micelle	*n.* 微胶粒
microbial	*adj.* 微生物的
microcrystalline cellulose	微晶纤维素
microenvironment	*n.* 微环境
microfluidic	*n.* 微流控
microgel	*n.* 微凝胶
microreactor	*n.* 微反应器
microRNAs	微小核糖核酸
migration	*n.* 迁移
miscellaneous	*adj.* 多方面的,其他的
moldy/mouldy	*adj.* 发霉的,霉烂的
molecular filtration media	分子过滤介质
momentum	*n.* 要素,动力
mono-bactam	单内酰胺化合物
monoclonal	*adj.* 单克隆的
monocyclic	*adj.* 单环的
monomer	*n.* 单体
mould	*n.* 霉,霉菌
moxifloxacin	*n.* 莫西沙星
mucous	*adj.* 黏液的,分泌黏液的 *n.* 黏膜
multidisciplinary	*adj.* 多学科的
munition	*n.* [复]军需(用)品,军火,弹药
mustard	*n.* 芥(禾),芥子气
mustard-gas	*n.* 芥子气
mutation	*n.* 体细胞突变
mycobiological	*adj.* 真菌生物学的
myelodysplastic	*adj.*骨髓增生异常
myeloid	*adj.* 脊椎的,骨髓的
myeloma	*n.* 骨髓瘤
myeloproliferative	*adj.* 骨髓增生

N

nalidixic acid 萘啶酸
nanogel *n.* 纳米凝胶
nanoparticle *n.* 纳米粒子
naphthyridone *n.* 萘啶酮
nasopharyngeal carcinoma 鼻咽癌
natural product 天然产物
nausea *n.* 反胃，恶心
neonate *n.* 新生儿
Nocardia *n.* 诺卡氏菌属，土壤细菌属
Nocardicin *n.* 诺卡杀菌素，诺卡地菌素
nonconformance *n.* 不符合项
non-hydrolyzable *adj.* 不可水解的
non-mutated gene 非变异基因
nonreproducible *adj.* 不能繁殖的，不能再生产的
Norfloxacin *n.* 诺氟沙星
nursing *n.* 哺乳
nutritional *adj.* 营养的

O

octanol *n.* 辛醇
ocular *n.* 眼睛；*adj.* 眼睛的，视觉的
oligonucleotide *n.* 寡(聚)核苷酸
oncology *n.* 肿瘤学
operating parameter 运行参数
operating range 运行范围
ophthalmic *adj.* 眼的；*n.* 眼药
oral *n.* 口试；*adj.* 口部的，口服的，口述的
orphan-drug *n.* 孤儿药
osmosis *n.* 渗透，渗透作用
osmotic *adj.* 渗透的；渗透性的
ovarian cancer 卵巢癌

P

package insert 包装说明书
Paclitaxel *n.* 紫杉醇
paintbrush *n.* 画笔，漆刷
pancreas *n.* 胰腺
parent compound 原型化合物，母体化合物
particle *n.* 颗粒
pathogen *n.* 病原体
pathological *adj.* 病理学的
pathophysiological *adj.* 病理生理
pediatric *adj.* 小儿科的
peer-reviewed 同行评议
pellet *n.* 药丸，小球
Penicillin *n.* 青霉素
penicillium notatum 青霉菌，特异青霉，点青霉
peptide *n.* 肽，缩氨酸
per *prep.* 每
petroleum ether 石油醚
pharmaceutical *adj.* 制药的，药学的；*n.* 药品，药剂
pharmacist *n.* 药剂师
pharmacodynamics *n.* 药效动力学，药效学
pharmacogenomics *n.* 药物基因组学
pharmacokinetic *n.* 药物代谢动力学
pharmacological *adj.* 药理学的，药物学的
pharmacophore 药效团
pharmacopoeia *n.* 药典
pharmacopoeial *adj.* 药典的
phenolic *adj.* 酚的
phenoxyacetic acid 苯氧基乙酸
phenylacetic acid 苯基乙酸
phosphoric *adj.* 磷的
photocarcinogenic *adj.* 光致癌的
physical examination 体格检查
pilot study 中试研究
pilot-plant 中试工厂，小规模试验性工厂
pipeline *n.* 管线
pipemidic acid 吡哌酸
pituitary *n.* 脑垂体
placeboarm *n.* 安慰剂对照组
plant *n.* 植物

plasma n. 血浆，淋巴液，等离子体
plasma concentration 血药浓度
plateletderived growth factor receptor 血小板衍生生长因子受体
pneumococcal adj. 肺炎双球菌的
pneumonia n. 肺炎
polydispersity index 多分散指数
polymerase n. 聚合酶
polymorphic forms 多晶型
polysaccharide n. 多糖，多聚糖
Pomalidomide n. 泊马度胺
postapproval n. 批准后
precipitate vt. 使沉淀；vi. 沉淀 n. 沉淀物
precision n. 精密，精确度，精确
preclinical adj. 临床前的
precursor n. 前体
pregnant adj. 妊娠的，怀孕的
premarket n. 上市前
prescribe v. 给医嘱，开处方
prescription n. 处方
prevalence n. 发(患)病率
principle 原则
prior medical history 既往病史
prior treatment history 既往治疗史
priority review 优先评审
process design 工艺设计
process parameters 工艺参数
process performance qualification 工艺性能确认
process qualification 工艺评价
process validation 工艺验证
processing limit 工艺限度
proliferation n. 增殖
prostate cancer 前列腺癌
protein n. 蛋白质
pseudomonas aeruginosa 铜绿假单胞菌
pulmonary adj. 肺的
pulsatile release 脉冲释放
pyrogen n. 致热质，热原

Q

qualify v. 鉴定
qualitative adj. 定性的
quantitative adj. 定量的
quinolones n. 喹诺酮类

R

radial adj. 径向的，(沿)半径的，放射的；n. 径向，光(射)线
radiographic tests (X)射线检测
radiolabeling n. 放射性标记
radiolabelled imaging agent 放射性同位素标记造影剂
radiopharmaceutical n. 放射性药物
radiosensitiser n. 放射增敏剂
radiotherapy n. 放射治疗
rare pediatric disease 罕见儿科疾病
rash n. 皮疹
raw material 起始原料，起始物料
rearrangement n. 重排
recrystallization n. 重结晶
recurrent adj. 复发的，周期性的
redox condition 氧化还原条件
reference standard 标准品
reflux n. 回流
refractory adj. 难治的
regime n. 方式，方法，规范
regiospecific reaction 区域专一性反应
registration n. 注册
release product 缓释制剂
replenishment n. (再)补给，充实，供给
reproducibility n. 可重复性
residence time 滞留时间
resolution n. 分辨率
respiratory adj. 呼吸(作用)的

response factor	响应因子
rigid	*n.* 刚性
rinse	*v.* 冲洗,漂洗; *n.* 漂清,冲洗
robust	*adj.* 强健的,健全的
rodent	*n.* 啮齿动物

S

salicylate	*n.* 水杨酸盐,水杨酸酯
salicylic acid	水杨酸
sampling plan	取样方案
Sardinian	*n.* 撒丁岛人,撒丁岛语; *adj.* 撒丁岛的,撒丁岛人的
scepticism	*n.* 怀疑论,怀疑主义
Schisandrin B	五味子素 B
semi-synthetic	半合成的
separatory funnel	分液漏斗
serum	*n.* 血浆
short circuit	短(捷)路,短接
short sequence	短序列
side-chain	侧链
side-effect	(药物的)副作用,不良反应
silicon dioxide	二氧化硅
slurry	*n.* 稀(泥, 沙)浆,悬浮体(液);*v.* 使变成泥浆
sodium	*n.* 钠
solubilizing	*adj.* 增容的
sonicate	*v.* 使(细胞,病毒等)经超声处理
sonochemical	*n.* 超声化学
soybean	*n.* 大豆,黄豆
specification	*n.* 说明书,规范,规格
spectrum/[复]spectra	*n.* 谱图
spore	*n.* 孢子; *vi.* 长孢子
staphylococcal	*adj.* 葡萄球菌的
staphylococcus/[复] *staphylococci*	n.葡萄球菌
Staphylococcus aureus	金黄色葡萄球菌
starch	*n.* 淀粉
statistical metrics	统计指标
steam bath	蒸气浴
stereochemistry	*n.* 立体化学
stereospecific reaction	立体专一性反应
sterile	*adj.* 无菌的,消毒的,不育的
steroid	*n.* 甾体
stoichiometry	*n.*化学计算,化学计量(法,学),理想配比法
stoppage	*n.* 停止,中止
stopper	*v.* 塞住; *n.* 塞子
stopwatch	*n.* 秒表,跑表
straightforward	*adj.*; *adv.*简单的(地),易懂的(地),坦率的(地)
streptokinase	*n.* 链球葡萄激酶
Streptomyces	*n.* 链霉菌素
Streptomycin	*n.* 链霉素
subcutaneous	*adj.* 皮下的
sublimation	*n.* 升华
substrate	*n.* 基质
suction funnel	吸入漏斗
sulfate	*v.* 硫酸化
sulphur	*n.*硫,硫黄
suspension	*n.* 混悬剂
swelling	*n.* 肿胀
synthetic lethality	合成致死,致死性

T

tannin	*n.* 丹宁,丹宁酸,鞣酸
target	*n.* 靶点
tendon	*n.* 肌腱
the blood-brain barrier	血脑屏障
therapeutical	*adj.* 治疗(学)的
therapy	*n.* 治疗
Thienamycin	*n.* 甲砜霉素
three neck round bottom flask	三口烧瓶
threshold	*n.* 阈值
thrombolytic	*n.* 溶栓剂

thyroidectomy　n. 甲状腺切除术
tissue-specific　组织特异性
titrimetrically　n. 滴定法
toluene　n. 甲苯
tonicity　n. 强壮,强健
toxicity　n. 毒性,毒力
toxicokinetics　n. 毒物代谢动力学
traceability　n. 追溯
transdermal　adj. 经皮的,透皮的
transformation　n. 变化,转化,转换,蜕变
transgenic　adj. 转基因的
tribute　n. 颂词,礼物,贡品
trypsin　n. 胰蛋白酶
tuberculosis　n. 结核(病),肺结核
tubular　adj. 管(状,形)的,有管的, 筒式的
tumour　肿瘤
tumor invasion　肿瘤侵袭
tumorigenic　adj. 致瘤的,致癌性,发生肿瘤的

U

ultrasound　n. 超声,超声波
unaltered　adj. 未修饰的,未被改变的
urinary　adj. (泌)尿的
urine　n. 尿液
utility　n. 公共设施

V

vaccine　n. 疫苗,牛痘疫苗
validity　n. 有效性
variation　n. 变更,变异
vehicle　n. 载体,传达媒介,赋形剂,运载工具
ventricular　adj. 心室的
verifying　n. 核实
versatility　n. 通用性
vesicant　adj. 起疱的(剂),腐烂性的
veterinary　n. 兽医;adj. 医牲畜的,兽医的
vial　n. 小瓶,小玻璃瓶;vt. 放……于小瓶中
virtual　n. 虚拟的
volatile　adj. 挥发性的 n. 挥发物
voltage　n. 电压
volume of production　产量
voucher　n. 凭证

W

Warfarin　n. 华法林
wavelength　n. 波长
wrap　vt. 包装

X

xylene　n. 二甲苯

Y

yeast　n. 酵母

References

1. BAI C K, YANG J J, CAO B, et al. Growth years and post-harvest processing methods have critical roles on the contents of medicinal active ingredients of *Scutellaria baicalensis*[J]. Industrial Crops & Products, 2020, 158, 112985.

2. EBABA S S, EDRADA R A, LIN W H, et al. Methods for isolation, purification and structural elucidation of bioactive secondary metabolites from marine invertebrates[J]. Nature Protocols, 2008, 3(12): 1820-1831.

3. MAYER A M, GLASER K B, CUEVAS C, et al. The odyssey of marine pharmaceuticals: a current pipeline perspective[J]. Trends in Pharmacological Sciences, 2010, 31(6): 255-265.

4. SABE V T, NTOMBELA T, JHAMBA L A, et al. Current trends in computer aided drug design and a highlight of drugs discovered via computational techniques: a review[J]. European Journal of Medicinal Chemistry, 2021, 224, 113705.

5. DAVE R, RANDHAWA G, KIM D, et al. Microgels and nanogels for the delivery of poorly water-soluble drugs[J]. Molecular Pharmaceutics 2022, 19(6): 1704-1721.

6. HU X Y, SONG Z, YANG Z W, et al. Cancer drug resistance related microRNAs: recent advances in detection methods[J]. Analyst, 2022, 147(12): 2615-2632.

7. BARNIEH F M, LOADMAN P M, FALCONER R A. Progress towards a clinically-successful ATR inhibitor for cancer therapy[J]. Current Research in Pharmacology and Drug Discovery, 2021, 2, 100017.

8. BISACCHI G S. Origins of the quinolone class of antibacterials: an expanded "discovery story"[J]. Journal of Medicinal Chemistry 2015, 58(12): 4874-4882.

9. LEVEQUE D, DELPEUCH A, GOURIEUX B. New Anticancer Agents: role of clinical pharmacy services[J]. Anticancer Research, 2014, 34(4): 1573-1578.

10. ICH Expert Working Group. ICH harmonised tripartite guideline: impurities in new drug substances Q3A(R2)[R], 2006.

11. ICH Expert Working Group. ICH harmonised tripartite guideline: guideline on the need for carcinogenicity studies of pharmaceuticals S1A[R], 1995.

12. U.S. Department of Health and Human Services, Food and Drug Administration, Center for Drug Evaluation and Research, et al. Process validation: general principles and practices[R], 2011.

13. U. S. Department of Health and Human Services, Food and Drug Administration, Center for Biologics Evaluation and Research, et al. Rare pediatric disease priority review vouchers[R], 2014.

14. U. S. Department of Health and Human Services, Food and Drug Administration, Center for Drug Evaluation and Research, et al. Determining the extent of safety data collection needed in late-stage premarket and postapproval clinical investigations[R], 2012.

15. Committee for Proprietary Medicinal Products. Points to consider on the clinical requirements of modified release products submitted as a line extension of an existing marketing authorization[EB/OL]. (2003-11-17)[2022-07-26]. https://www.ema.europa.eu/en/documents/scientific-guideline/points-consider-clinical-requirements-modified-release-products-submitted-line-extension-existing_en.pdf.

16. European Medicines Agency. Guideline on the pharmacokinetic and clinical evaluation of modified release dosage forms[R]. 2014.

17. MONGA K, MYRICK O. Digital pill that 'talks' to your smartphone approved for first time[EB/OL]. (2017-11-15)[2023-12-08]. https://abcnews.go.com/Health/digital-pill-talks-smartphone-approved-time/story?id=51161456.

18. The United States Pharmacoperial Convention. The United States pharmacopeia: general chapters[S]. Rockville: The United States Pharmacoperial Convention, 2020.

19. Novartis Pharmaceuticals Corporation. GLEEVEC-imatinib mesylate tablet [EB/OL]. (2022-05)[2022-07-25]. https://nctr-crs.fda.gov/fdalabel/ui/spl-summaries/criteria/108276.

20. DING Z C, CHEN S H, ZHAO B P, et al. Substituted 2-hydrogen-pyrazole derivative serving as anticancer drug: EP3269715 [P]. 2018-01-17.

21. PATEL O P S, BETECK R M, LEGOABE L J. Exploration of artemisinin derivatives and synthetic peroxides in antimalarial drug discovery research [J]. European Journal of Medicinal Chemistry: Chimie Therapeutique, 2021, 213, 113193.

22. SCOTT L J. Apatinib: a review in advanced gastric cancer and other advanced cancers[J]. Drugs, 2018, 78(7): 747-758.

23. ICH Expert Working Group. Good manufacturing practice guide for active pharmaceutical ingredients Q7[R]. 2020.

24. SALDIVAR-GONZALEZ F I, ALDAS-BULOS V D, MEDINA-FRANCO J L, et al. Natural product drug discovery in the artificial intelligence era[J]. Chemical Science, 2022, 13(6): 1526-1546.

25. BANAKAR V V, SABNIS S S, GOGATE P R, et al. Ultrasound assisted continuous processing in microreactors with focus on crystallization and chemical synthesis: a critical review[J]. Chemical Engineering Research and Design, 2022, 182: 273-289.

26. LIU X Y, QIN Y. Industrial total synthesis of natural medicines[J]. Natural Product Report, 2023, 40: 1694-1700.

27. World Health Organization. The use of stems in the selection of International Nonproprietary Names (INN) for pharmaceutical substances 2018 (StemBook2018)[R/OL]. 2018.

28. 尤启冬.药物化学[M].8 版.北京:人民卫生出版社,2016.

29. 吴达俊, 庄思永.制药工程专业英语[M].北京: 化学工业出版社, 2018.

30. 刘宇红.化学化工专业英语[M].北京: 中国轻工业出版社,2006.

31. 张永勤, 刘福胜.生物与制药工程专业英语[M].北京: 化学工业出版社, 2020.

32. 吴月凤, 赵林静, 秦莉萍.英汉构词法对比探析[J].英语广场: 学术研究, 2019, 6:50-51.

33. 胡婷.石油化工英语的构词特征与翻译技巧[J].粘接, 2019, 40(8):168-170.

34. 李羚玮.构词法在英语词汇教学中的应用研究[J].湖南工业职业技术学院学报, 2015, 15(4):81-83.

35. 姚媛.英汉互译:分析与实践[M].南京:南京大学出版社,2017.

36. 刘丽红, 蒋翠.英语翻译技巧与原则研究[M].长春:吉林出版集团股份有限公司, 2018.

37. 周季特,庄德君.写译[M].哈尔滨:哈尔滨工业大学出版社,1998.

38. 詹蓓.浅说英文字母的象形工程作用[J].安徽工业大学学报(社会科学版),2004, 21(2):74-75, 91.

39. 张仲景.伤寒论[M].罗希文,译.北京: 新世界出版社,2007.

40. 佚名.黄帝内经[M].罗希文,译.北京: 中国中医药出版社,2009.